Assessment Guide

Module H

HOLT McDOUGAL

HOUGHTON MIFFLIN HARCOURT

Acknowledgements for Covers

Cover Photo Credits

Light echo in space (bg) ©NASA/ESA/Hubble Heritage Team (STScI/AURA); *pacific wheel* (l) ©Geoffrey George/Getty Images; *snowboarder* (cl) ©Jonathan Nourok/Photographer's Choice/Getty Images; *water droplet* (cr) ©L. Clarke/Corbis; *molecular structure* (r) ©Stockbyte/Getty Images

Copyright © by Houghton Mifflin Harcourt Publishing Company

All rights reserved. No part of this work may be reproduced or transmitted in any form or by any means, electronic or mechanical, including photocopying or recording, or by any information storage and retrieval system, without the prior written permission of the copyright owner unless such copying is expressly permitted by federal copyright law. Requests for permission to make copies of any part of the work should be addressed to Houghton Mifflin Harcourt Publishing Company, Attn: Contracts, Copyrights, and Licensing, 9400 South Park Center Loop, Orlando, Florida 32819.

Printed in the U.S.A.

ISBN 978-0-547-59343-2

22 2266 24 23
4500869540   A B C D E F G

If you have received these materials as examination copies free of charge, Houghton Mifflin Harcourt Publishing Company retains title to the materials and they may not be resold. Resale of examination copies is strictly prohibited.

Possession of this publication in print format does not entitle users to convert this publication, or any portion of it, into electronic format.

Contents

INTRODUCTION
Overview vi
Formative Assessment viii
    Assessing Prior Knowledge viii
    Embedded Assessment ix
Summative Assessment xi
    In the Student Edition xi
    In This Assessment Guide xi
    Performance-Based Assessment xiii
    Portfolio Assessment, Guidelines xiv

ASSESSMENT TOOLS
Alternative Assessment Presentation Guidelines xv
Alternative Assessment Rubrics
    Tic-Tac-Toe xvi
    Mix and Match xvii
    Take Your Pick xviii
    Choose Your Meal xix
    Points of View xx
    Climb the Pyramid xxi
    Climb the Ladder xxii
Classroom Observation Checklist xxiii
Lab and Activity Evaluation xxiv
Portfolio Planning Worksheet xxv
Portfolio Evaluation Checklist xxvi

Unit 1 Matter
Unit 1 Matter: Pretest 1
Lesson 1 Quiz: Introduction to Matter 3
Lesson 2 Quiz: Properties of Matter 4
Lesson 3 Quiz: Physical and Chemical Changes 5
Lesson 4 Quiz: Pure Substances and Mixtures 6
Lesson 5 Quiz: States of Matter 7
Lesson 6 Quiz: Changes of State 8
Lesson 1 Alternative Assessment: Introduction to Matter 9
Lesson 2 Alternative Assessment: Properties of Matter 10
Lesson 3 Alternative Assessment: Physical and Chemical Changes 11
Lesson 4 Alternative Assessment: Pure Substances and Mixtures 12
Lesson 5 Alternative Assessment: States of Matter 13
Lesson 6 Alternative Assessment: Changes of State 14
Performance-Based Assessment: Teacher Edition 15
Performance-Based Assessment: Student Edition 16
Unit 1 Review 18
Unit 1 Test A 24
Unit 1 Test B 31

Unit 2 Energy
Unit 2 Energy: Pretest 38
Lesson 1 Quiz: Introduction to Energy 40
Lesson 2 Quiz: Temperature 41
Lesson 3 Quiz: Thermal Energy and Heat 42
Lesson 4 Quiz: Effects of Energy Transfer 43
Lesson 1 Alternative Assessment: Introduction to Energy 44
Lesson 2 Alternative Assessment: Temperature 45
Lesson 3 Alternative Assessment: Thermal Energy and Heat 46
Lesson 4 Alternative Assessment: Effects of Energy Transfer 47
Performance-Based Assessment: Teacher Edition 48
Performance-Based Assessment: Student Edition 49
Unit 2 Review 51
Unit 2 Test A 55
Unit 2 Test B 61

Unit 3 Atoms and the Periodic Table
Unit 3 Atoms and the Periodic Table: Pretest 67
Lesson 1 Quiz: The Atom 69
Lesson 2 Quiz: The Periodic Table 70
Lesson 3 Quiz: Electrons and Chemical Bonding 71
Lesson 4 Quiz: Ionic, Covalent, and Metallic Bonding 72
Lesson 1 Alternative Assessment: The Atom 73
Lesson 2 Alternative Assessment: The Periodic Table 74
Lesson 3 Alternative Assessment: Electrons and Chemical Bonding 75
Lesson 4 Alternative Assessment: Ionic, Covalent, and Metallic Bonding ... 76
Performance-Based Assessment: Teacher Edition 77
Performance-Based Assessment: Student Edition 78
Unit 3 Review 80
Unit 3 Test A 84
Unit 3 Test B 90

Unit 4 Interactions of Matter
Unit 4 Interactions of Matter: Pretest 96
Lesson 1 Quiz: Chemical Reactions 98
Lesson 2 Quiz: Organic Chemistry 99
Lesson 3 Quiz: Nuclear Reactions 100
Lesson 1 Alternative Assessment: Chemical Reactions 101
Lesson 2 Alternative Assessment: Organic Chemistry 102
Lesson 3 Alternative Assessment: Nuclear Reactions 103
Performance-Based Assessment: Teacher Edition 104
Performance-Based Assessment: Student Edition 105
Unit 4 Review 107
Unit 4 Test A 111
Unit 4 Test B 117

Unit 5 Solutions, Acids, and Bases
    Unit 5 Solutions, Acids, and Bases: Pretest 123
    Lesson 1 Quiz: Solutions 125
    Lesson 2 Quiz: Acids, Bases, and Salts 126
    Lesson 3 Quiz: Measuring pH 127
    Lesson 1 Alternative Assessment: Solutions 128
    Lesson 2 Alternative Assessment: Acids, Bases, and Salts 129
    Lesson 3 Alternative Assessment: Measuring pH 130
    Performance-Based Assessment: Teacher Edition 131
    Performance-Based Assessment: Student Edition 132
    Unit 5 Review 134
    Unit 5 Test A 138
    Unit 5 Test B 144

End-of-Module Test 150

Answer Sheet 160

Answer Key 161

INTRODUCTION
Overview

ScienceFusion provides parallel instructional paths for teaching important science content. You may choose to use the print path, the digital path, or a combination of the two. The quizzes, tests, and other resources in this Assessment Guide may be used with either path.

The *ScienceFusion* assessment options are intended to give you maximum flexibility in assessing what your students know and what they can do. The program's formative and summative assessment categories reflect the understanding that assessment is a learning opportunity for students, and that students must periodically demonstrate mastery of content in cumulative tests.

All *ScienceFusion* tests are available—and editable—in ExamView and online at thinkcentral.com. You can customize a quiz or test for your classroom in many ways:

- adding or deleting items
- adjusting for cognitive complexity, Bloom's taxonomy level, or other measures of difficulty
- changing the sequence of items
- changing the item formats
- editing the question itself

All of these changes, except the last, can be made without invalidating the content correlation of the item.

This Assessment Guide is your directory to assessment in *ScienceFusion*. In it you'll find copymasters for Lesson Quizzes, Unit Tests, Unit Reviews, Performance-Based Assessments Alternative Assessments, and End-of-Module Tests; answers and explanations of answers; rubrics; a bubble-style answer sheet; and suggestions for assessing student progress using performance, portfolio, and other forms of integrated assessment.

You will also find additional assessment prompts and ideas throughout the program, as indicated on the chart that follows.

Assessment in ScienceFusion Program

	Student Editions	Teacher Edition	Assessment Guide	Digital Lessons	Online Resources at thinkcentral.com	ExamView Test Generator
Formative Assessment						
Assessing Prior Knowledge						
Engage Your Brain	X					
Unit Pretest			X		X	X
Embedded Assessment						
Active Reading Questions	X					
Interactivities	X					
Probing Questions		X				
Formative Assessment		X				
Classroom Discussions		X				
Common Misconceptions		X				
Learning Alerts		X				
Embedded Questions and Tasks				X		
Student Self-Assessments				X		
Digital Lesson Quiz				X		
When used primarily for teaching						
Lesson Review	X	X				
Lesson Quiz			X		X	X
Alternative Assessment			X		X	
Performance-Based Assessment			X			
Portfolio Assessment, guidelines			X			
Summative Assessment						
End of Lessons						
Visual Summary	X	X				
Lesson Quiz			X		X	X
Alternative Assessment		X	X		X	
Rubrics			X		X	
End of Units						
Unit Review	X		X		X	X
Answers		X	X		X	
Test Doctor Answer Explanations		X	X			X
Unit Test A (on level)			X		X	X
Unit Test B (below level)			X		X	X
End of Module						
End-of-Module Test			X		X	X

Formative Assessment
Assessing Prior Knowledge

Frequently in this program, you'll find suggestions for assessing what your students already know before they begin studying a new lesson. These activities help you warm up the class, focus minds, and activate students' prior knowledge.

In This Assessment Guide
Each of the units begins with a Unit Pretest consisting of multiple-choice questions that assess prior and prerequisite knowledge. Use the Pretest to get a snapshot of the class and help you organize your pre-teaching.

In the Student Edition
Engage Your Brain Simple, interactive warm-up tasks get students thinking, and remind them of what they may already know about the lesson topics.

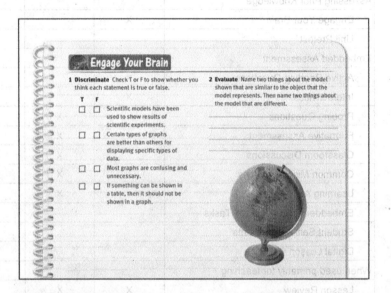

Active Reading Questions Students first see the lesson vocabulary on the opening page, where they are challenged to show what they know about the terms. Multiple exposures to the key terms throughout the lesson lead to mastery.

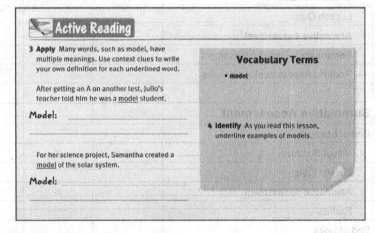

In the Teacher Edition
Opening Your Lesson At the start of each TE lesson Opening Your Lesson suggests questions and activities that help you assess prerequisite and prior knowledge.

Embedded Assessment

Once you're into the lesson, you'll continue to find suggestions, prompts, and resources for ongoing assessment.

Student Edition

Active Reading Questions and Interactivities Frequent questions and interactive prompts are embedded in the text, where they give students instant feedback on their comprehension. They ask students to respond in different ways, such as writing, drawing, and annotating the text. The variety of skills and response types helps all students succeed, and keeps them interested.

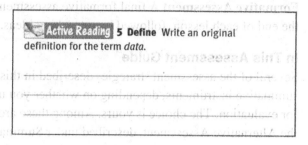

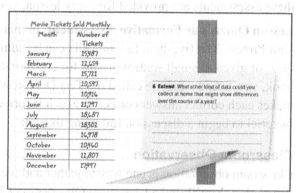

In the Teacher Edition

Probing Questions Probing questions appear in the point-of-use teaching suggestions. These questions are labeled to show the degree of independent inquiry they require. The three levels of inquiry—Directed, Guided, and Independent—give students experience that builds toward independent analysis.

Classroom Discussions Discussion is a natural opportunity to gauge how well students have absorbed the material, and to assess any misconceptions or gaps in their understanding. Students also learn from each other in this informal exchange. Classroom discussion ideas appear throughout the lesson in the Teacher Edition.

Tips for Classroom Discussions
- Allow students plenty of time to reflect and formulate their answers.
- Call upon students you sense have something to add but who haven't spoken.
- At the same time, allow reluctant students not to speak unless they choose to.
- Encourage students to respond to each other as well as to you.

Misconceptions and Learning Alerts The Teacher Background pages at the start of a unit describe common misconceptions and identify the lessons in which the misconceptions can be addressed. Strategies for addressing the misconceptions appear in the point-of-use teaching notes. Additional Learning Alerts help you introduce and assess challenging topics.

Formative Assessment A final formative assessment strategy appears on the Evaluate page at the end of each lesson, followed by reteaching ideas.

In This Assessment Guide
Several of the assessment strategies described in this book can be used either as formative or as summative instruments, depending on whether you use them primarily for teaching or primarily for evaluation. The choice is yours. Among these are the Lesson Quizzes, described here, and the Alternative Assessment, described under Summative Assessment, next. Because both of these assessments are provided for every lesson, you could use them both at different times.

Lesson Quizzes as Formative Assessment In this book, Lesson Quizzes in a unit follow the Unit Pretest. The five-item Lesson Quiz can be administered as a written test, used as an oral quiz, or given to small student groups for collaboration. In the Answer Key at the end of this book, you'll find a feature called the Test Doctor, which provides a brief explanation of what makes each correct answer correct and each incorrect answer incorrect. Use this explanatory material to begin a discussion following the quiz.

Classroom Observation
Classroom observation is one way to gather and record information that can lead to improved instruction. You'll find a Classroom Observation Checklist in Assessment Tools, following the Introduction.

Tips for Classroom Observation
- Don't try to see and record everything at once. Instead, identify specific skills you will observe in a session.
- Don't try to observe everyone at once. Focus on a few students at a time.
- Repeat observations at different times in order to identify patterns. This practice helps you validate or correct your impressions from a single time.
- Use the checklist as is or modify it to suit your students and your instruction. Fill in student names across the top and write the date next to the skills you are observing on a particular day.
- Keep the checklist, add to it, and consult it periodically for hints about strengths, weaknesses, and developments of particular students and of the class.
- Use your own system of ratings or the simple number code on the checklist. When you have not seen enough to give a rating, leave the space blank.

Summative Assessment

In the Student Edition

Visual Summary and Lesson Review
Interactive summaries help students synthesize lesson material, and the Lesson Review provides a variety of questions focusing on vocabulary, key concepts, and critical thinking.

Unit Reviews
Each unit in the Student Edition is followed by a Unit Review, also available in this Assessment Guide. These tests include the item types commonly found on the statewide assessments. You may want to use these tests to review unit content right away or at any time later in the year to help students prepare for the statewide assessment. If you wish to give students practice in filling in a machine-scorable answer sheet, use the bubble-type answer sheet at the start of the Answer Key.

In This Assessment Guide

Alternative Assessments
Every lesson has an Alternative Assessment worksheet, which is previewed in the Teacher Edition on the Evaluate page of the lesson. The activities on these worksheets assess student comprehension of core content, while at the same time offering a variety of options for students with various abilities, learning styles, and interests. The activities require students to produce a tangible product or to give a presentation that demonstrates their understanding of skills and concepts.

Tips for Alternative Assessment

- The structure of these worksheets allows for differentiation in topic, difficulty level, and activity type/learner preferences.

- Each worksheet has a variety of items for students and teachers to choose from.

- The items may relate to the entire lesson content or to just one or two key topics. Encourage students to select items so that they will hit most key topics in a lesson.

- Share the rubrics and Presentation Guidelines with students so they understand the expectations for these assignments. You could have them fill in a rubric with their name and activity choices at the same time they choose their assignments, and then submit the rubric with their presentation or assignment.

Grading Alternative Assessments

Each type of Alternative Assessment worksheet has a rubric for easy grading.

- The rubrics focus mostly on content comprehension, but also take into account presentation.
- The Answer Key describes the expected content mastery for each Alternative Assessment.
- Separate Presentation Guidelines describe the attributes of successful written work, posters and displays, oral presentations, and multimedia presentations.
- Each rubric has space to record your reasons for deducting points, such as content errors or particular presentation flaws.
- If you wish to change the focus of an Alternative Assessment worksheet, you can adjust the point values for the rubric.

The Presentation Guidelines and the rubrics follow the Introduction. The Answer Key appears at the end of the book.

Unit Tests A and B

This Assessment Guide contains leveled tests for each unit.

- The A-level tests are for students who typically perform below grade level.
- The B-level tests are intended for students whose performance is on grade level.

Both versions of the test address the unit content with a mixture of item types, including multiple choice, short response, and extended response. Both levels contains items of low, medium, and high cognitive complexity, though level B contains more items of higher complexity. A few items appear in both of the tests as a means of assuring parallel content coverage. If you need a higher-level test, you can easily assemble one from the lesson assessment banks in ExamView or online at thinkcentral.com. All items in the banks are tagged with five different measures of difficulty as well as standards and key terms.

End-of-Module Test

The final test in this Assessment Guide is the End-of-Module Review. This is a long-form, multiple-choice test in the style of the statewide assessments. An Answer Sheet appears with the review.

Performance-Based Assessment

Performance-Based Assessment involves a hands-on activity in which students demonstrate their skills and thought processes. Each Performance-Based Assessment includes a page of teacher-focused information and a general rubric for scoring. In addition to the Performance-Based Assessment provided for each unit, you can use many of the labs in the program as the basis for performance assessment.

Tips for Performance Assessment

- Prepare materials and stations so that all students have the same tasks. You may want to administer performance assessments to different groups over time.

- Provide clear expectations, including the measures on which students will be evaluated. You may invite them to help you formulate or modify the rubric.

- Assist students as needed, but avoid supplying answers to those who can handle the work on their own.

- Don't be hurried. Allow students enough time to do their best work.

Developing or Modifying a Rubric

Developing a rubric for a performance task involves three basic steps:

1. Identify the inquiry skills that are taught in the lesson and that students must perform to complete the task successfully and identify the understanding of content that is also required. Many of the skills may be found in the Lab and Activity Evaluation later in this guide.

2. Determine which skills and understandings of content are involved in each step.

3. Decide what you will look for to confirm that the student has acquired each skill and understanding you identified.

Portfolio Assessment, Guidelines

A portfolio is a showcase for student work, a place where many types of assignments, projects, reports and data sheets can be collected. The work samples in the collection provide snapshots of the student's efforts over time, and taken together they reveal the student's growth, attitudes, and understanding better than other types of assessment. Portfolio assessment involves meeting with each student to discuss the work and to set goals for future performance. In contrast with formal assessments, portfolio assessments have these advantages:

1. They give students a voice in the assessment process.
2. They foster reflection, self-monitoring, and self-evaluation.
3. They provide a comprehensive picture of a student's progress.

Tips for Portfolio Assessment

- Make a basic plan. Decide how many work samples will be included in the portfolios and what period of time they represent.

- Explain the portfolio and its use. Describe the portfolio an artist might put together, showing his or her best or most representative work, as part of an application for school or a job. The student's portfolio is based on this model.

- Together with your class decide on the required work samples that everyone's portfolio will contain.

- Explain that the students will choose additional samples of their work to include. Have students remember how their skills and understanding have grown over the period covered by the portfolio, and review their work with this in mind. The best pieces to choose may not be the longest or neatest.

- Give students the Portfolio Planning Worksheet found in Assessment Tools. Have students record their reasoning as they make their selections and assemble their portfolios.

- Share with students the Portfolio Evaluation Checklist, also found in Assessment Tools, and explain how you will evaluate the contents of their portfolios.

- Use the portfolios for conferences, grading, and planning. Give students the option of taking their portfolios home to share.

ASSESSMENT TOOLS
Alternative Assessment Presentation Guidelines

The following guidelines can be used as a starting point for evaluating student presentation of alternative assessments. For each category, use only the criteria that are relevant for the particular format you are evaluating; some criteria will not apply for certain formats.

Written Work
- Matches the assignment in format (essay, journal entry, newspaper report, etc.)
- Begins with a clear statement of the topic and purpose
- Provides information that is essential to the reader's understanding
- Supporting details are precise, related to the topic, and effective
- Follows a logical pattern of organization
- Uses transitions between ideas
- When appropriate, uses diagrams or other visuals
- Correct spelling, capitalization, and punctuation
- Correct grammar and usage
- Varied sentence structures
- Neat and legible

Posters and Displays
- Matches the assignment in format (brochure, poster, storyboard, etc.)
- Topic is well researched and quality information is presented
- Poster communicates an obvious, overall message
- Posters have large titles and the message, or purpose, is obvious
- Images are big, clear, and convey important information
- More important ideas and items are given more space and presented with larger images or text
- Colors are used for a purpose, such as to link words and images
- Sequence of presentation is easy to follow because of visual cues, such as arrows, letters, or numbers
- Artistic elements are appropriate and add to the overall presentation
- Text is neat
- Captions and labels have correct spelling, capitalization, and punctuation

Oral Presentations
- Matches the assignment in format (speech, news report, etc.)
- Presentation is delivered well, and enthusiasm is shown for topic
- Words are clearly pronounced and can easily be heard
- Information is presented in a logical, interesting sequence that the audience can follow
- Visual aids are relative to content, very neat, and artistic
- Often makes eye contact with audience
- Listens carefully to questions from the audience and responds accurately
- Stands straight, facing the audience
- Uses movements appropriate to the presentation; does not fidget
- Covers the topic well in the time allowed
- Gives enough information to clarify the topic, but does not include irrelevant details

Multimedia Presentations
- Topic is well researched, and essential information is presented
- The product shows evidence of an original and inventive approach
- The presentation conveys an obvious, overall message
- Contains all the required media elements, such as text, graphics, sounds, videos, and animations
- Fonts and formatting are used appropriately to emphasize words; color is used appropriately to enhance the fonts
- Sequence of presentation is logical and/or the navigation is easy and understandable
- Artistic elements are appropriate and add to the overall presentation
- The combination of multimedia elements with words and ideas produces an effective presentation
- Written elements have correct spelling, capitalization, and punctuation

Alternative Assessment Rubric – Tic-Tac-Toe

Worksheet Title: _____

Student Name: _____

Date: _____

Add the titles of each activity chosen to the chart below.

	Content (0-3 points)	Presentation (0-2 points)	Points Sum
Choice 1: _____			
Points			
Reason for missing points			
Choice 2: _____			
Points			
Reason for missing points			
Choice 3: _____			
Points			
Reason for missing points			
		Total Points (of 15 maximum)	

Assessment Tools

© Houghton Mifflin Harcourt Publishing Company

Alternative Assessment Rubric – Mix and Match

Worksheet Title: _____

Student Name: _____

Date: _____

Add the column choices to the chart below.

	Content *(0-3 points)*	**Presentation** *(0-2 points)*	**Points Sum**
Information Source from Column A: _____			
Topics Chosen for Column B: _____			
Presentation Format from Column C: _____			
Points			
Reason for missing points			
		Total Points (of 5 maximum)	

Assessment Tools
© Houghton Mifflin Harcourt Publishing Company

Module H • Assessment Guide

Alternative Assessment Rubric – Take Your Pick

Worksheet Title: _____

Student Name: _____

Date: _____

Add the titles of each activity chosen to the chart below.

	Content	Presentation	Points Sum
2-point item: 5-point item: 8-point item:	*(0-1.5 points)* *(0-4 points)* *(0-6 points)*	*(0-0.5 point)* *(0-1 point)* *(0-2 points)*	
Choice 1: _____			
Points			
Reason for missing points			
Choice 2: _____			
Points			
Reason for missing points			
		Total Points (of 10 maximum)	

Assessment Tools
© Houghton Mifflin Harcourt Publishing Company

Module H • Assessment Guide

Alternative Assessment Rubric – Choose Your Meal

Worksheet Title: _____

Student Name: _____

Date: _____

Add the titles of each activity chosen to the chart below.

Appetizer, side dish, or dessert: Main Dish	**Content** (0-3 points) (0-6 points)	**Presentation** (0-2) points (0-4 points)	**Points Sum**
Appetizer: _____			
Points			
Reason for missing points			
Side Dish: _____			
Points			
Reason for missing points			
Main Dish: _____			
Points			
Reason for missing points			
Dessert: _____			
Points			
Reason for missing points			
		Total Points (of 25 maximum)	

Assessment Tools
© Houghton Mifflin Harcourt Publishing Company

Module H • Assessment Guide

Alternative Assessment Rubric – Points of View

Worksheet Title: _____

Student Name: _____

Date: _____

Add the titles of group's assignment to the chart below.

	Content (0-4 points)	**Presentation** (0-1 points)	**Points Sum**
Point of View:			
Points			
Reason for missing points			
		Total Points (of 5 maximum)	

Assessment Tools
© Houghton Mifflin Harcourt Publishing Company

Module H • Assessment Guide

Alternative Assessment Rubric – Climb the Pyramid

Worksheet Title: _____

Student Name: _____

Date: _____

Add the titles of each activity chosen to the chart below.

	Content (0-3 points)	**Presentation** (0-2 points)	**Points Sum**
Choice from bottom row: _____			
Points			
Reason for missing points			
Choice from middle row: _____			
Points			
Reason for missing points			
Top row: _____			
Points			
Reason for missing points			
		Total Points (of 15 maximum)	

Alternative Assessment Rubric – Climb the Ladder

Worksheet Title: _____

Student Name: _____

Date: _____

Add the titles of each activity chosen to the chart below.

	Content (0-3 points)	**Presentation** (0-2 points)	**Points Sum**
Choice 1 (top rung): _____			
Points			
Reason for missing points			
Choice 2 (middle rung): _____			
Points			
Reason for missing points			
Choice 3 (bottom rung): _____			
Points			
Reason for missing points			
		Total Points (of 15 maximum)	

Date _____

Classroom Observation Checklist

Rating Scale			
☐ 3	Outstanding	☐ 1	Needs Improvement
☐ 2	Satisfactory	☐	Not Enough Opportunity to Observe

Names of Students / Inquiry Skills										
Observe										
Compare										
Classify/Order										
Gather, Record, Display, or Interpret Data										
Use Numbers										
Communicate										
Plan and Conduct Simple Investigations										
Measure										
Predict										
Infer										
Draw Conclusions										
Use Time/Space Relationships										
Hypothesize										
Formulate or Use Models										
Identify and Control Variables										
Experiment										

Assessment Tools
© Houghton Mifflin Harcourt Publishing Company

Module H • Assessment Guide

Name _____ Date _____

Lab and Activity Evaluation

Lab and Activity Evaluation

Circle the appropriate number for each criterion. Then add up the circled numbers in each column and record the sum in the subtotals row at the bottom. Add up these subtotals to get the total score.

Graded by _____ Total _____ /100

Behavior	Completely	Mostly	Partially	Poorly
Follows lab procedures carefully and fully	10–9	8–7–6	5–4–3	2–1–0
Wears the required safety equipment and displays knowledge of safety procedures and hazards	10–9	8–7–6	5–4–3	2–1–0
Uses laboratory time productively and stays on task	10–9	8–7–6	5–4–3	2–1–0
Behavior	**Completely**	**Mostly**	**Partially**	**Poorly**
Uses tools, equipment, and materials properly	10–9	8–7–6	5–4–3	2–1–0
Makes quantitative observations carefully, with precision and accuracy	10–9	8–7–6	5–4–3	2–1–0
Uses the appropriate SI units to collect quantitative data	10–9	8–7–6	5–4–3	2–1–0
Records accurate qualitative data during the investigation	10–9	8–7–6	5–4–3	2–1–0
Records measurements and observations in clearly organized tables that have appropriate headings and units	10–9	8–7–6	5–4–3	2–1–0
Works well with partners	10–9	8–7–6	5–4–3	2–1–0
Efficiently and properly solves any minor problems that might occur with materials or procedures	10–9	8–7–6	5–4–3	2–1–0
Subtotals:				

Comments

_____

_____

_____

_____

Assessment Tools
© Houghton Mifflin Harcourt Publishing Company

Name _____ Date _____ Date _____

Portfolio Planning Worksheet

My Science Portfolio

What Is in My Portfolio	Why I Chose It
1.	
2.	
3.	
4.	
5.	
6.	
7.	

I organized my Science Portfolio this way because _____

_____

_____

_____

_____

Assessment Tools
© Houghton Mifflin Harcourt Publishing Company

Module H • Assessment Guide

Name _____ Date _____

Portfolio Evaluation Checklist

Portfolio Evaluation Checklist

Aspects of Science Literacy	Evidence of Growth
1. **Understands science concepts** (Animals, Plants; Earth's Land, Air, Water; Space; Weather; Matter, Motion, Energy)	_____ _____ _____
2. **Uses inquiry skills** (observes, compares, classifies, gathers/interprets data, communicates, measures, experiments, infers, predicts, draws conclusions)	_____ _____ _____
3. **Thinks critically** (analyzes, synthesizes, evaluates, applies ideas effectively, solves problems)	_____ _____ _____
4. **Displays traits/attitudes of a scientist** (is curious, questioning, persistent, precise, creative, enthusiastic; uses science materials carefully; is concerned for environment)	_____ _____ _____

Summary of Portfolio Assessment

For This Review			Since Last Review		
Excellent	Good	Fair	Improving	About the Same	Not as Good

Name _____ Date _____

Unit 1

Pretest

Matter

Choose the letter of the best answer.

1. The illustration below shows a rectangular solid.

   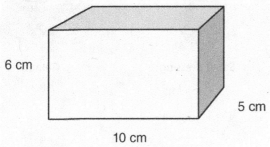

   What is the volume of this solid?

   A. 21 cm³
   B. 30 cm²
   C. 60 cm²
   D. 300 cm³

2. All matter has physical and chemical properties. These properties can be used to identify the type of matter. Which of these statements describes a chemical property?

   A. A particular substance evaporates at 30 °C.
   B. A 2-ft.-long metal bar has a mass of only 176 g.
   C. A certain heavy metal turns to a liquid at room temperature.
   D. A metal is added to a beaker of water, and the beaker explodes.

3. David found that water can be created in a lab by burning hydrogen gas in air. He concluded that water is not a compound because only hydrogen was used to form water. What is wrong with David's conclusion?

   A. A compound does contain only one type of element.
   B. Hydrogen is made up of two different types of atoms.
   C. Water was not the product formed when he burned hydrogen.
   D. The hydrogen combined with oxygen from the air to form water.

4. Which process represents a chemical change?

   A. A lake freezes over into ice.
   B. A metal bar is rolled into a flat sheet.
   C. Vinegar bubbles when baking soda is added.
   D. Sand, water, and salt combine to form a mixture.

5. The diagram below shows how the temperature of water changes as the water changes states.

   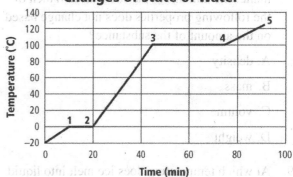

   Between which points does water boil?

   A. 1 to 2
   B. 2 to 3
   C. 3 to 4
   D. 4 to 5

6. Trini adds 10 g of baking soda to 100 g of vinegar. The mixture begins to bubble. When the bubbling stops, Trini finds the mass of the resulting mixture. She determines its mass is 105 g. Why has the mass changed?

   A. A gas has formed and left the mixture.
   B. Vinegar evaporated during the experiment.
   C. Mixtures always are less massive than their parts.
   D. Mass was destroyed when vinegar reacted with baking soda.

Name _____ Date _____ Date _____

Unit 1

7. Marissa blows up balloons for a party. She decides how big or small to make each balloon. Why does the air she blows into each balloon take up all the space inside the balloon?

   A. The air particles are easily able to slide past one another.

   B. Air is a gas and so fills its container, the balloon, completely.

   C. The air particles increase in size when they have more space.

   D. There is a strong attraction between the air particles and the balloon.

8. Some properties are the same in a substance no matter the amount of the substance. Which of the following properties does not change based on the amount of the substance?

   A. density
   B. mass
   C. volume
   D. weight

9. At which temperature does ice melt into liquid water?

   A. 0°C
   B. 32°C
   C. 100°C
   D. 212°C

10. The four items below were part of a dinner. Each item is a mixture.

Salad dressing
A

Gelatin
B

Whipped cream
C

Apple juice
D

Which of these mixtures is a suspension?

A. A
B. B
C. C
D. D

Pretest
© Houghton Mifflin Harcourt Publishing Company

Module H • Assessment Guide

Name _____ Date _____

Unit 1 Lesson 1

Lesson Quiz

Introduction to Matter

Choose the letter of the best answer.

1. Which statement is true of all matter?

   A. It has mass.

   B. It can be seen.

   C. It exists only as a solid.

   D. It maintains its shape and size.

2. A metal coin has certain properties that can be measured. Which property of a coin is different on the moon than it is on Earth?

   A. mass

   B. weight

   C. volume

   D. density

3. What is the volume of a rectangular solid that is 40 centimeters long, 10 centimeters wide, and 5 centimeters high?

   A. 400 cm^3

   B. 500 cm^3

   C. 1,000 cm^3

   D. 2,000 cm^3

4. Which phrase describes the mass of an ice cube?

   A. the total amount of water in the ice cube

   B. an estimation of the weight of the ice cube

   C. the amount of space that the ice cube occupies

   D. the product found by multiplying its length, width, and height

5. The table lists the densities of some common materials at 20°C.

Material	Density (g/cm^3)
gasoline	0.70
mercury	13.6
milk	1.03
water	0.998

   If a scientist has 10 grams of each material, which material has the greatest volume?

   A. milk

   B. water

   C. mercury

   D. gasoline

Lesson Quiz

Module H • Assessment Guide

Name _____ Date _____

Unit 1 Lesson 2

Lesson Quiz

Properties of Matter

Choose the letter of the best answer.

1. Hydrogen gas (H_2) can be found in trace amounts in Earth's atmosphere. Which of these statements describes a physical property of hydrogen?

   A. Hydrogen is found in acids.

   B. Hydrogen gas is highly flammable.

   C. Hydrogen reacts with oxygen to form water.

   D. Hydrogen gas is less dense than oxygen gas.

2. Which of these is a chemical property of a sheet of paper?

   A. The paper can be burned.

   B. The paper can be crumpled.

   C. The paper does not attract a magnet.

   D. The paper does not conduct electricity.

3. Which of these choices is a physical property that does not change when the size of the sample changes?

   A. mass

   B. volume

   C. density

   D. flammability

4. The pictures below show four objects—a paper clip, a pair of scissors, a needle, and a horseshoe. Assume that each object is made of the same metal.

   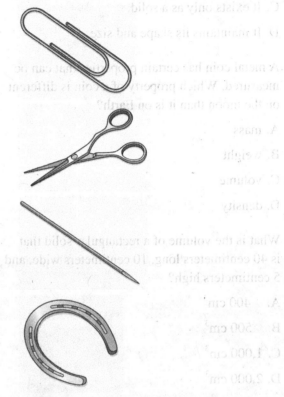

   Which of these physical properties is not similar in all four of these objects?

   A. mass

   B. magnetism

   C. specific heat

   D. electrical conductivity

5. Which of these statements describes a chemical property of an object?

   A. The object is white in color.

   B. The object has a powdery texture.

   C. The object's density is 2.11 g/cm^3.

   D. The object reacts with acid to form water.

Name _____ Date _____ Date _____

Unit 1 Lesson 3

Lesson Quiz

Physical and Chemical Changes

Choose the letter of the best answer.

1. Which process is an example of a physical change?

    A. Carrots are cut into small pieces and mixed into a salad.

    B. A peanut butter sandwich is eaten and broken down by enzymes in the stomach.

    C. Sodium metal and chlorine gas are combined to form sodium chloride, or table salt.

    D. Sodium metal and water are combined to form a basic compound and a flammable gas.

2. Which process is an example of a chemical change?

    A. an iron nail rusting

    B. bath water cooling while you take a bath

    C. a piece of metal being heated until it expands

    D. a glass window breaking when hit with a baseball

3. When paper is burned, the mass of the remaining ash is less than the mass of the original paper. Which statement best explains this result?

    A. The ash has less volume than the paper.

    B. Some of the matter is destroyed during the reaction.

    C. The mass of the ash cannot be accurately determined.

    D. Some of the products of the reaction were given off as a gas.

4. There are several differences between chemical and physical changes. Which process is an example of a chemical change?

    A. steam rising from a boiling pot of soup

    B. a metal railing rusting in damp weather

    C. alcohol evaporating from a cotton swab

    D. a piece of wood shrinking as it dries out

5. Marco tears a piece of notebook paper into smaller pieces, as shown below.

    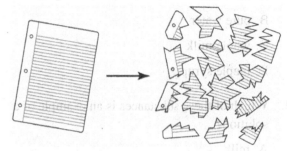

    Tearing paper into pieces is an example of what kind of change?

    A. a change in mass

    B. a physical change

    C. a chemical change

    D. a change in energy

Name _____ Date _____

Unit 1 Lesson 4

Lesson Quiz

Pure Substances and Mixtures

Choose the letter of the best answer.

1. What type of substance is always made up of a single type of atom?

   A. mixture

   B. element

   C. molecule

   D. compound

2. Which of these common substances is a homogeneous mixture?

   A. table salt

   B. pure water

   C. whole milk

   D. maple syrup

3. Which of these substances is an example of a solution?

   A. milk

   B. brass

   C. mercury

   D. concrete

4. Reactant A and reactant B undergo a chemical reaction to form product C.

   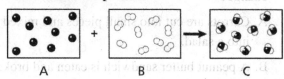

   What type of substance is product C?

   A. an atom

   B. a mixture

   C. an element

   D. a compound

5. Which of these substances is a compound?

   A. carbon

   B. chlorine

   C. uranium

   D. ammonia

Lesson Quiz
© Houghton Mifflin Harcourt Publishing Company

Module H • Assessment Guide

Name _____ Date _____

Unit 1 Lesson 5
Lesson Quiz

States of Matter

Choose the letter of the best answer.

1. Ana is creating a model to show atoms of solid bromine, liquid bromine, and gaseous bromine. How should her three models differ?

   A. The size of the atoms should vary depending on the state.

   B. The mass of the atoms should vary depending on the state.

   C. The motion of the atoms should vary depending on the state.

   D. The identity of the atoms should vary depending on the state.

2. Which state of matter will take both the volume and shape of the container that holds it?

   A. gas

   B. ice

   C. liquid

   D. solid

3. Frost forms when water vapor changes directly to ice in a process called deposition. If you were to model the water particles before and after deposition, how would they compare?

   A. Before deposition, the particles vibrate in place; after deposition, they slide by each other.

   B. Before deposition, the particles slide by each other; after deposition, they vibrate in place.

   C. Before deposition, the particles vibrate in place; after deposition, they move quickly in all directions.

   D. Before deposition, the particles move quickly in all directions; after deposition, they vibrate in place.

4. Why does water that is frozen in an ice cube tray stay in the shape of a cube when it is taken out of the tray?

   A. The water particles become locked in place.

   B. The water particles stop moving completely.

   C. The water particles grow bigger to fill the space.

   D. The water particles can only slip past one another.

5. The diagram illustrates particles in three states of matter.

   1   2   3

   What do each of the states of matter have in common?

   A. The particles are locked into position.

   B. The particles in each are in constant motion.

   C. The particles take the shape of their containers.

   D. The particles have the same volume in each container.

Name _____ Date _____

Unit 1 Lesson 6

Lesson Quiz

Changes of State

Choose the letter of the best answer.

1. Dry ice is solid carbon dioxide. At room temperature, it changes directly into a gas. A model of which of the following would describe this change?

    A. Evaporation results in a reduction of the mass of carbon dioxide.

    B. Freezing occurs due to a decrease in the kinetic energy of the particles.

    C. Sublimation occurs due to an increase in the kinetic energy of the particles.

    D. Deposition causes the particles of the carbon dioxide gas to lock into place.

2. A student measured the mass of ice in a glass container with a tight lid. He allowed the ice to melt, and then found the mass of the container, its lid, and its contents again. What conclusion could he draw based on the masses he measured?

    A. No matter how tight the lid is, some mass is lost when a solid melts.

    B. The liquid formed from a melted solid has less mass than the solid has.

    C. The liquid formed from a melted solid has the same mass as the solid has.

    D. The liquid formed from a melted solid has a greater mass than the solid has.

3. Nora notices water droplets on the grass in the morning. It did not rain during the night. Which statement is true about this change of state?

    A. Mass was added to the water particles, resulting in deposition.

    B. Energy was added to the water particles, resulting in evaporation.

    C. Mass was removed from the water particles, resulting in sublimation.

    D. Energy was removed from the water particles, resulting in condensation.

4. Which changes of state result in a decrease in the kinetic energy of the particles?

    A. sublimation, melting, boiling

    B. melting, freezing, evaporation

    C. sublimation, deposition, melting

    D. deposition, freezing, condensation

5. The following illustration shows three different states of a substance.

    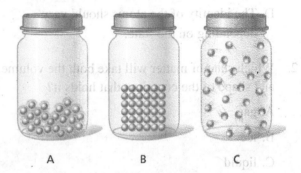

    A   B   C

    Which of the following happens when the substance in jar B changes state to the substance in jar A?

    A. The particles vibrate faster.

    B. The mass of the substance increases.

    C. The identity of the substance changes.

    D. The particles expand to fill their container.

Name _____ Date _____

Unit 1 Lesson 1

Alternative Assessment

Introduction to Matter

Points of View: *Mass, Volume, Density*
Your class will work together to show what you've learned about matter from several different viewpoints.

1. Work in groups as assigned by your teacher. Each group will be assigned to one or two viewpoints.

2. Complete your assignment, and present your perspective to the class.

 Examples List and describe ways that people can measure volume, mass, and density. Give examples of objects whose volumes, masses, and densities can be measured using each method.

 Illustrations Imagine your friend is measuring the volume of his or her home for a school project. Your friend's home is a rectangle that contains six rectangle-shaped rooms and one rectangle-shaped hallway. Diagram several ways the rooms could be arranged. Show several ways your friend could measure the total interior volume of the home.

 Analysis Explain how you could find the mass and weight of an object on Earth, and the mass and weight of the same object on the moon. How would mass and weight be similar and different in these two places?

 Observations Find two similarly sized objects in your classroom. First, write a guess about which object will have a larger volume, mass, and density. Then use displacement or a formula to determine the volumes of the objects. Compare the masses. Think about the densities. Then describe the volumes, masses, and densities of the two objects.

 Calculations Imagine your neighbors are installing a swimming pool that is a perfect rectangle. They want to know the volume of their pool, but they do not know how to measure it. The pool is 3.0 meters wide, 5.0 meters in length, and 2 meters deep. Determine the volume of the swimming pool. If they only fill the pool to a depth of 1.5 meters, what is the volume of water that it will take to fill it? The neighbors have two balls for the pool that are the same size, but one floats and one sinks. Compare the mass, volume, and density of the two balls.

Name _____ Date _____

Unit 1 Lesson 2

Alternative Assessment

Properties of Matter

Climb the Pyramid: *Identifying Physical and Chemical Properties*
Complete the activities to show what you've learned about physical and chemical properties of matter.

1. Work on your own, with a partner, or with a small group.
2. Choose one item from each layer of the pyramid. Check your choices.
3. Have your teacher approve your plan.
4. Submit or present your results.

___ **What Is It?**
Imagine you find a bowl filled with clear liquid in your kitchen. List three liquids it might be. Describe how you could use the liquid's physical and chemical properties to determine which of the three it is.

___ **Trading Properties**
Design trading cards that identify and describe physical properties. On your cards, write the names of at least five physical properties (such as density, texture, mass, volume, color, state). Then explain the properties and give examples.

___ **Be the Scientist!**
Design an experiment in which you test at least one chemical property of a common object or item. Include a hypothesis about the object's chemical properties and include a method for testing the object.

___ **Watch the Reaction**
Mix baking soda and vinegar. Identify the type of property shown. Then talk about what your observations tell you about the properties of baking soda and vinegar.

___ **Advising a Friend**
Imagine your friend tries to test the chemical properties of wood. However, the friend confuses a physical property with a chemical one! Tell what the friend did, and explain why the observation will not identify a chemical property. Then suggest what should be done.

___ **Showing Properties**
Choose an object in the classroom. Then list different physical properties of the object. Write a paragraph or make a labeled diagram that includes at least four different physical properties of the object.

Name _____ Date _____

Unit 1 Lesson 3

Alternative Assessment

Physical and Chemical Changes

Take Your Pick: *What a Change!*
Complete the activities to show what you've learned about physical and chemical changes.

1. Work on your own, with a partner, or with a small group.
2. Choose items below for a total of 10 points. Check your choices.
3. Have your teacher approve your plan.
4. Submit or present your results.

2 Points

_____ **Everyday Science** Describe a way the law of conservation of mass can be applied to daily life.

_____ **Chemical Observations** Think about a common chemical change. Write a brief journal entry in which you describe how the chemical change occurs and the results of the change.

5 Points

_____ **Did It Disappear?** Imagine you left a cup filled with a powder-based drink on the counter. A few days later, only crystals were left in the cup. Perform a skit in which you explain what happened. Use the law of conservation of mass in your explanation.

_____ **Charting Chemical Change** Imagine your friend's backyard has an iron chair, a fire pit filled with wood, and a copper downspout. Map your friend's backyard. Indicate three chemical changes that could occur there.

_____ **Picturing Change** Design a poster or collage that shows both physical and chemical changes occurring to matter. Label and describe the physical and chemical changes.

8 Points

_____ **Water, Water Everywhere** Imagine you are on a walk with a friend. During your walk, you feel water vapor rising from a grate, see ice on the sidewalk and water drops falling from a roof. Your friend asks how water changes from one state to the other. Write or perform a dialogue in which you answer the question.

_____ **Cooking up Change** Make a multimedia presentation in which you identify physical and chemical changes that can occur while cooking.

_____ **Going in Reverse** Both physical and chemical changes can be reversed. Write an essay about reversing chemical and physical changes. List examples of physical and chemical changes that can be reversed. Which type of change seems more common?

Name _____ Date _____

Unit 1 Lesson 4

Alternative Assessment

Pure Substances and Mixtures

Choose Your Meal: *Matter Menu*
Mix and match ideas to show what you've learned about pure substances and mixtures.

1. Work on your own, with a partner, or with a small group.
2. Choose one item from each section of the menu, with an optional dessert. Check your choices.
3. Have your teacher approve your plan.
4. Submit or present your results.

Appetizers

_____ **Salty and Sweet** Table salt and table sugar are common compounds. Do some quick research to identify the elements that make up each compound.

_____ **Make a Model** Design a model to explain the differences between an element and a compound. Share your model with the class.

_____ **Table It** Make a table or other graphic organizer that identifies different types of mixtures and their characteristics. Identify two food products that are examples of each type of mixture.

Main Dish

_____ **Parts of Your Entree** You are given a mixture of sand, iron nails, and salt water. Devise a plan for separating this mixture. Identify how you will separate each part of the mixture and the physical properties that allow you to do this.

Side Dishes

_____ **Recipe for Success** Write a recipe card for a food you enjoy that is a mixture. Choose a food that includes at least three ingredients. Explain how the ingredients are put together to make the food.

_____ **Metals, Nonmetals, and Metalloids** Elements are often classified as metals, nonmetals, and metalloids. Draw a description wheel for each type of element. Include on the spokes of the wheel examples of metals, nonmetals, and metalloids, and the properties of each.

Desserts (optional)

_____ **Concept Mapping** Develop a concept map that uses the following terms: *matter, element, compound, mixture, solution, suspension,* and *colloid*.

_____ **Shake Before Use** The words "Shake Before Use" on a product label usually indicate that a substance is a suspension. Identify three food products that could contain this label, and explain why the substance needs to be shaken before use.

Name _____ Date _____

Unit 1 Lesson 5

States of Matter

Alternative Assessment

Points of View: *Particles in Motion*
Your class will work together to show what you've learned about states of matter from several different viewpoints.

1. Work in groups as assigned by your teacher. Each group will be assigned to one or two viewpoints.

2. Complete your assignment, and present your perspective to the class.

Examples Produce a multimedia presentation in which you show the same substance in more than one state. For each state, show how the molecules move.

Illustrations Design a poster that shows the molecules in solids, liquids, and gases. In each image, indicate how close the molecules are to one another and indicate how they move.

Analysis Solids, liquids, and gases have differences in their shapes and volumes. Think about the properties of cookie or bread dough. Describe whether you think dough behaves more like a solid or a liquid, and why.

Observations Find two online animations that show how water looks as a liquid, solid, and a gas. Compare and critique the two animations.

Details Imagine your school day has been canceled because of an ice storm. Present a skit in which you describe what you see out your window at different points in the day. In your skit, explain how water changes shape and volume at different times of the day.

Name _____ Date _____

Unit 1 Lesson 6
Alternative Assessment

Changes of State

Points of View: *State of Mind*
Your class will work together to show what you've learned about changes of state from several different viewpoints.

1. Work in groups as assigned by your teacher. Each group will be assigned to one or two viewpoints.

2. Complete your assignment, and present your perspective to the class.

 Terms Write a paragraph discussing changes of state using the words *change, directly, movement, melting point, intermediate,* and any other words or phrases that pertain to changes of state.

 Details Describe why the mass of a substance does not change when its state changes. Describe what characteristics do change.

 Illustrations Draw a picture of the particles of a material as it changes state. Then draw another picture of the particles in a material changing to a different state. Label the pictures with short descriptions of what is happening as well as the names for these changes of state.

 Examples Give examples of changes of state of matter.

 Analysis Compare the movement of particles in solids, liquids, and gases. How are changes in the movement of these particles related to changes of state?

The Density of Metals

Unit 1

Performance-Based Assessment Teacher Notes

Purpose Students will measure the mass and volume of copper and zinc to calculate their densities, and will use this information to determine which coins are made of copper and which are made of zinc.

Time Period 45-60 minutes

Preparation Make sure you use graduated cylinders with the smallest, most precise increments possible. The diameter of the graduated cylinder should be wide enough so that a quarter can easily fall in and out. Cover each activity station with a tarp or a drop cloth. Equip each activity station with the necessary materials.

Pennies minted before 1982 were made of 95 percent copper and 5 percent zinc. From 1982 and thereafter, pennies were made of copper-plated zinc. The pennies used in this activity should all be minted before 1982. Nickels are 75 percent copper and 25 percent nickel. Dimes and quarters are manufactured with a pure copper core and a face that is 75 percent copper and 25 percent nickel. Although none of the coins will have precisely the same density as the pure metals, students should be able to distinguish whether the coins consist mostly of copper or zinc. This will ensure consistency in the measurement of density.

Safety Tips Spilled water is a slipping hazard. Clean up water spills immediately. Students should immediately notify the teacher if a graduated cylinder breaks or if a student cuts himself or herself. Have a sharps container nearby in case of glass breakage.

Teaching Strategies This activity works best in groups of 4–5 students. Before the activity, review how to calculate density. Show students how to correctly read a water level in a graduated cylinder. Remind students that in order to get an accurate reading, they must take the reading at the bottom of the meniscus.

Scoring Rubric

Possible points	Performance indicators
0-30	Appropriate use of materials and equipment
0-40	Making observations and testing hypotheses
0-30	Drawing conclusions

Name _____ Date _____

Unit 1

Performance-Based Assessment

The Density of Metals

Objective
You have learned that the density of an object is related to its mass and volume. Every metal has a unique density, which can be used to identify it. In this activity, you will calculate the densities of two metals, and use this information to find out what metals coins are made of.

Know the Score!
As you work through this activity, keep in mind that you will be earning a grade for the following:
- how well you work with the materials and equipment (30%)
- how well you make observations and test the hypothesis (40%)
- how accurately you identify test objects (30%)

Materials and Equipment
- balance
- calculator
- copper sample
- cylinder, graduated
- dimes (10)
- nickels (10)
- pennies (10)
- water, 1 L
- zinc sample

Safety Information
- Clean up water spills immediately; spilled water is a slipping hazard.

Procedure
1. Compare the luster and appearance of the coins with the luster and appearance of the copper and zinc samples.

   _____
   _____
   _____

2. Each coin is made of a mixture of metals. Which metals do you think the coins are mostly made of? Explain your predictions.

   _____
   _____
   _____
   _____

Name _____ Date _____   Unit 1

3. Place the copper sample on the balance. Measure and record the mass in the table below.

Metal sample	Mass of metal (g)	Starting volume (mL)	Ending volume (mL)	Volume of metal (mL)	Density of metal (g/mL)
Copper					
Zinc					
Pennies					
Nickels					
Dimes					

4. Pour water into the graduated cylinder until it is half full. Record the starting volume of the water in the table.

5. Place the copper sample in the cylinder. Record the ending volume of the water in the table.

6. Subtract the starting volume from the ending volume to find the volume of the metal. Record this value in the table.

7. Repeat steps 3–6 with the zinc sample. Record your measurements.

8. Repeat steps 3–6 with the 10 pennies. Record your measurements. Continue to repeat steps 3–6 until all the sets of coins have been tested.

Analysis

9. Calculate the density of each metal by dividing the mass of the metal by the volume of the metal. Record the density in the table above. Show your work below.

10. Compare the densities of the coins with the densities of the known metals. Based on the densities, what are the coins mostly made of? Support your answer.

_____

_____

_____

Name _____ Date _____

Unit 1

Unit Review

Unit 1: Matter

Vocabulary
Check the box to show whether each statement is true or false.

T	F	
☐	☐	1. <u>Matter</u> is anything that has mass and takes up space.
☐	☐	2. An <u>element</u> is a substance that is made up of one type of atom.
☐	☐	3. <u>Evaporation</u> is the change of state from a gas to a liquid.
☐	☐	4. A <u>solid</u> has a definite volume and shape.
☐	☐	5. A <u>physical property</u> can be measured without changing the identity of the substance.

Key Concepts
Read each question below, and circle the best answer.

6. One chemical property that can be measured in a substance is its reactivity with water. What is another chemical property?

   A. density

   B. flammability

   C. malleability

   D. solubility

7. Matter is made up of particles. Which of the following statements is true about these particles?

   A. The particles that make up solids do not move.

   B. The particles that make up liquids do not move.

   C. The particles that make up all matter are constantly in motion.

   D. Only the particles that make up gases are constantly in motion.

Unit Review
© Houghton Mifflin Harcourt Publishing Company

Module H • Assessment Guide

Name _____ Date _____

Unit 1

8. Two balloons are inflated to an equal volume. Balloon 2 is placed in the freezer for 20 minutes.

Balloon 1    Balloon 2

Why would freezing a balloon produce the results shown in Balloon 2?

A. Increased kinetic energy decreases the attraction between particles inside the balloon.

B. Increased kinetic energy increases the attraction between particles inside the balloon.

C. Decreased kinetic energy decreases the attraction between particles inside the balloon.

D. Decreased kinetic energy increases the attraction between particles inside the balloon.

9. Which of the following statements describes a liquid?

A. A liquid has both a definite shape and a definite volume.

B. A liquid has neither a definite shape nor a definite volume.

C. A liquid has a definite shape but not a definite volume.

D. A liquid has a definite volume but not a definite shape.

10. A water molecule is made up of one oxygen atom and two hydrogen atoms. Why is water considered a pure substance?

A. Water can be broken down by physical means.

B. Water can be combined with other substances by physical means.

C. Each water molecule is identical.

D. Water molecules are made up of different types of atoms.

Name _____ Date _____

Unit 1

11. A beaker containing a certain substance has heat applied to it. The particles that make up the substance begin to move farther apart from each other.

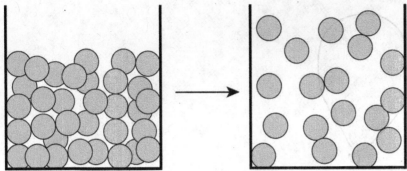

What change of state could be occurring to the substance in the beaker?

A. The substance is changing from a gas to a liquid.

B. The substance is changing from a gas to a solid.

C. The substance is changing from a liquid to a solid.

D. The substance is changing from a liquid to a gas.

12. The law of conservation of mass states that mass cannot be created or destroyed. To what type of change does this law apply?

A. physical changes only

B. chemical changes only

C. both physical and chemical changes

D. only mass that is not undergoing change

13. A beaker containing ice and water is placed on a warm hotplate. Will the ice in the beaker undergo a physical or chemical change?

A. a physical change because it will change state

B. a chemical change because it will change state

C. a physical change because it will form a new substance

D. a chemical change because it will form a new substance

14. What is the boiling point of water?

A. 0° C

B. 32° C

C. 100° C

D. 212° C

Name _____ Date _____

Unit 1

15. A rock is dropped into a graduated cylinder filled with 35 mL of water.

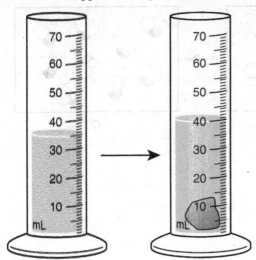

What is the volume of the rock? (Hint: 1 mL water = 1 cm³)

A. 40 cm³

B. 14 cm³

C. 5 cm³

D. 35 cm³

16. The instrument below is used to measure an object.

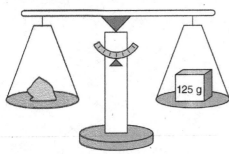

What is the instrument measuring?

A. gravity

B. weight

C. density

D. mass

Unit Review 21 Module H • Assessment Guide
© Houghton Mifflin Harcourt Publishing Company

Name _____ Date _____   Unit 1

17. The diagram below shows a chemical reaction.

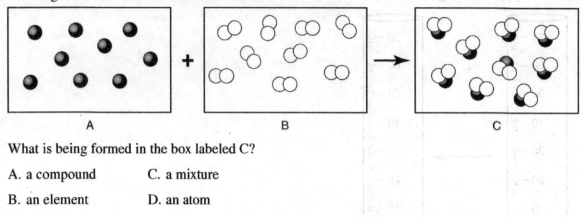

What is being formed in the box labeled C?

A. a compound   C. a mixture

B. an element   D. an atom

Critical Thinking
Answer the following questions in the space provided.

18. In the space below, sketch the particles in a solution, a suspension, and a colloid.

Give an example of a solution, a suspension, and a colloid.

Solution: _____

Suspension: _____

Colloid: _____

19. A sample liquid is heated in a closed container until it changes to a gas. What happens to the size of the particles in the sample?

_____

What happens to the number of particles in the sample?

_____

What happens to the average speed of the particles?

_____

Name _____ Date _____

Unit 1

20. Describe the difference between a chemical change and a physical change.
_____
_____

What are three examples of physical changes?
_____
_____
_____

What are three signs that a chemical change has taken place?
_____
_____
_____

How does temperature affect chemical changes?
_____
_____

Connect ESSENTIAL QUESTIONS

Lessons 1 and 2

Answer the following question in the space provided.

21. An unknown substance has a volume of 2 cm³ and a mass of 38.6 grams. What is the density of the sample? _____

Material	Density (g/cm³)
water	1.0
aluminum	2.7
iron	7.9
silver	10.5
gold	19.3

Use the chart above to find the identity of the unknown sample: _____

List three other physical properties that could be used to identify this sample.
_____
_____
_____

Unit Review 23 Module H • Assessment Guide

Name _____ Date _____

Unit 1

Matter

Unit Test A

Key Concepts

Choose the letter of the best answer.

1. Alisa placed a small seashell on the pan of a balance. What other tool does she need to determine the density of the seashell?

   A. scale

   B. calculator

   C. thermometer

   D. graduated cylinder

2. Magnetism, solubility, and malleability are physical properties of matter. What makes these properties different from chemical properties?

   A. Physical properties relate to elements rather than compounds.

   B. Physical properties appear only after a chemical change occurs.

   C. Physical properties can be observed without attempting to change the identity of the substance.

   D. Physical properties describe elements in the solid state rather than in the liquid or gas state.

3. The diagram below models a change that occurs in matter.

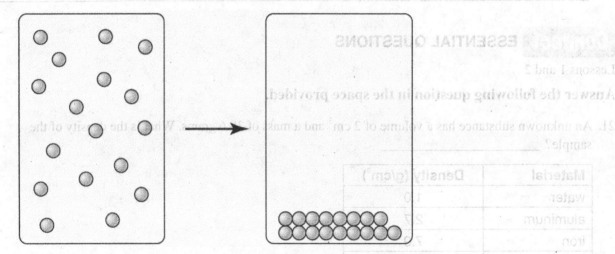

   Based on what you know about changes in matter, explain why the diagram shows a physical change rather than a chemical change.

   A. The particles get smaller.

   B. The volume of the matter decreases.

   C. The composition of the matter stays the same.

   D. The particles change into particles of different substances.

Unit Test A
© Houghton Mifflin Harcourt Publishing Company

Module H • Assessment Guide

Name _____ Date _____

Unit 1

4. A child is upset because his ice cream is melting. He thinks he now has less dessert. Which of the following explanations correctly states why the child is incorrect?

   A. He actually has more, not less, ice cream.

   B. No mass is lost during a change of state.

   C. The mass will increase if he freezes the melted ice cream.

   D. Mass is lost only during certain changes of state, such as freezing.

5. Josie partially filled a graduated cylinder with water. She then dropped a rock into the water. The illustration below shows what happened to the level of the water inside the graduated cylinder.

   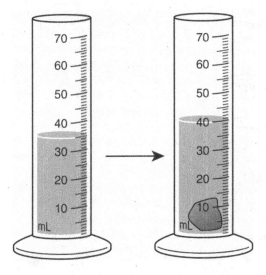

   Which property of matter is Josie measuring?

   A. mass

   B. length

   C. weight

   D. volume

6. Which of the following is a way in which elements and compounds are similar?

   A. Elements and compounds are both pure substances.

   B. Elements and compounds are both listed on the periodic table.

   C. Elements and compounds are both made up of different kinds of atoms.

   D. Elements and compounds can both be broken down by physical changes.

Name _____ Date _____

Unit 1

7. You are shining a light through liquids A and B, which are both mixtures. You can see the light pass through A, but B looks cloudy and you see tiny particles floating in it.

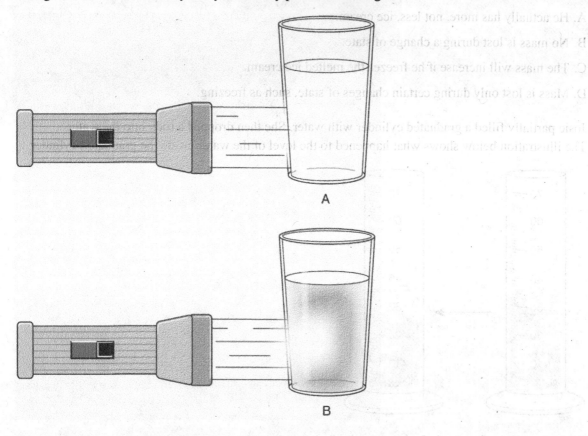

Which of these statements best describes the liquids?

A. Both the liquids are solutions.

B. Both the liquids are suspensions.

C. A is a solution, and B is a suspension.

D. A is a suspension, and B is a solution.

8. Which process is an example of a physical change?

A. burning

B. rusting

C. flattening

D. decomposing

9. The diagram below shows a magnet near a pile of particles of iron and sulfur. The magnet attracts the iron, separating it from the mixture.

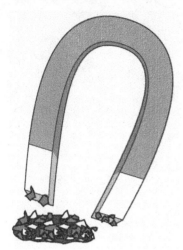

Based on the diagram, which statement is true?

A. The parts of a mixture keep their own properties.

B. The elements in a compound keep their own properties.

C. The properties of a mixture are different from the properties of its parts.

D. The properties of a compound are different from the properties of its elements.

10. Chemical changes result in new substances, but physical changes do not. Which process is an example of a chemical change?

A. chopping a tree

B. cooking a steak

C. heating a cup of tea

D. drying clothes in the dryer

11. Which types of changes observe the law of conservation of mass?

A. only changes of state

B. only physical changes

C. only chemical changes

D. physical changes and chemical changes

12. The diagram below shows a Venn diagram to compare the properties of solids, liquids, and gases.

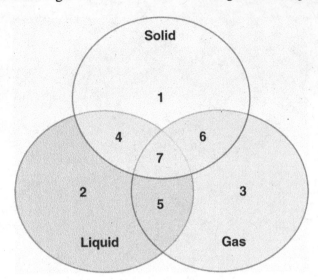

What property, which only solids have, is in position 1?

A. definite shape

B. definite volume

C. particles that do not move

D. particles are always in motion

13. What happens to the temperature of a sample of matter as it changes from a liquid to a gas?

A. Temperature increases.

B. Temperature decreases.

C. Temperature remains constant.

D. Temperature first decreases, and then increases.

14. On a hot day, a puddle dries up. Which statement describing this event is true?

A. The water in the puddle boils and becomes a gas.

B. The water in the puddle evaporates into water vapor.

C. The water in the puddle condenses and becomes part of air.

D. The water in the puddle changes into particles of oxygen in the air.

Name _____ Date _____ Unit 1

15. The two beakers shown below contain pure water.

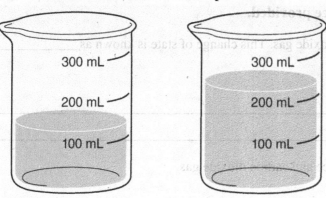

Which of the following is a chemical property of the water in both beakers?

A. The water is colorless.

B. The water freezes at 0 °C.

C. The water reacts with some metals.

D. The water has a total volume of about 400 mL.

Critical Thinking
Answer the following questions in the space provided.

16. A class wants to use marbles in a clear shoe box with a lid to model the particles in a solid, a liquid, and a gas.

    Describe the motion of particles in each state of matter.

    _____

    _____

    Describe how you would model the motion of the particles in each state.

    _____

    _____

    _____

Name _____ Date _____ Unit 1

Extended Response
Answer the following questions in the space provided.

17. Solid dry ice changes directly into carbon dioxide gas. This change of state is known as _____

What happens during this change of state?

_____

_____

Compare the motion of the particles in dry ice and carbon dioxide gas.

_____

_____

Compare the original mass of dry ice with the mass of carbon dioxide gas that forms. Explain how you arrived at your answer.

_____

_____

Name _____ Date _____

Matter

Key Concepts

Choose the letter of the best answer.

1. Amin had two metal cubes of identical sizes. He placed cube A on one side of a pan balance and cube B on the other side of the balance. The pan that held cube B was now lower than the pan that held cube A. What conclusion can Amin draw about the two cubes?

   A. The masses of the two cubes are the same.

   B. The volumes of the two cubes are different.

   C. The weights of the two cubes are the same.

   D. The densities of the two cubes are different.

2. Which of these choices is an example of a physical property?

   A. the tarnishing of silverware

   B. the texture of a piece of chocolate

   C. the effect of acid rain on automobiles

   D. the combustion of gasoline in a car engine

3. The diagram below shows a type of change in matter at the particle level. Study it closely.

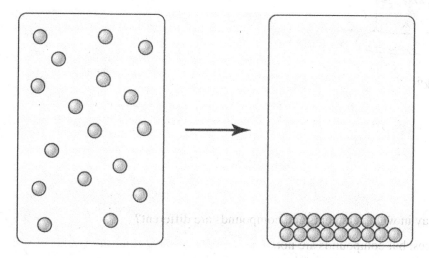

   Based on what you know about changes in matter, which type of change does the diagram show?

   A. a change in mass

   B. a physical change

   C. a chemical change

   D. a change in reactivity

Name _____ Date _____ Unit 1

4. An ice cube melts and then the liquid water evaporates. Which statement is true about the mass of water that goes through each change of state?

   A. The mass of the water in each state is the same because mass is conserved.

   B. The mass of water increases as the water's volume increases with each change of state.

   C. The mass of water vapor is less than the mass of ice because some water is lost during each change of state.

   D. The mass of water vapor is less than the mass of ice because the density of water vapor is less than the density of ice.

5. Herman partially filled a graduated cylinder with water. He then dropped a rock into the water. The illustration below shows what happened to the level of the water inside the graduated cylinder.

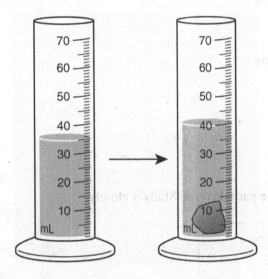

   What is the volume of the rock?

   A. 5 mL

   B. 10 mL

   C. 35 mL

   D. 40 mL

6. Which of the following is a way in which elements and compounds are different?

   A. Elements are pure substances, but compounds are not.

   B. Compounds can be broken down by physical changes, but elements cannot.

   C. Elements are made of identical atoms, whereas compounds are made of identical molecules.

   D. A compound contains many different kinds of elements, whereas an element contains many different kinds of atoms.

Name _____ Date _____

Unit 1

7. You know that one of these containers has a mixture in it and one does not. You can only shine a light through them to determine which one is which.

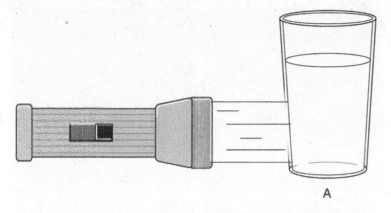

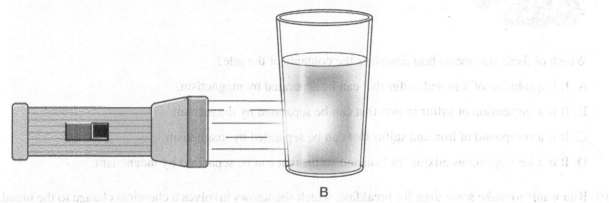

What substance is most likely to be in container A?

A. water

B. gelatin

C. apple juice

D. mayonnaise

8. Which process is an example of a physical change?

A. Water turns to steam when boiled over a Bunsen burner.

B. Carbon combines with oxygen to form carbon dioxide gas.

C. Water breaks down into hydrogen and oxygen gases over time.

D. Limestone breaks down into lime and carbon dioxide when heated

Name _____ Date _____

Unit 1

9. A magnet was placed near a pile that contained both iron and sulfur. The magnet was moved gradually closer to the pile. As it neared the pile, the magnet started attracting small pieces of iron from the pile.

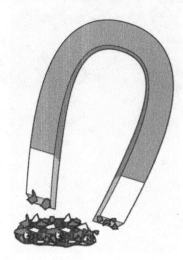

Which of these statements best describes the contents of the pile?

A. It is a solution of iron and sulfur that can be separated by magnetism.

B. It is a suspension of sulfur in iron that can be separated by magnetism.

C. It is a compound of iron and sulfur that can be separated by magnetism.

D. It is a heterogeneous mixture of iron and sulfur that can be separated by magnetism.

10. Rita wants to make some toast for breakfast, which she knows involves a chemical change to the bread. She puts a slice of bread in the toaster, but, after 10 minutes, she notices that the sides of the bread are black. What caused this chemical change to go too far?

A. the type of bread

B. the size of the bread

C. a decrease in temperature

D. an increase in temperature

11. Marisol finds the total mass of a sample of baking soda, a balloon, and a glass container with vinegar in it. She adds the baking soda to the vinegar and immediately places the balloon over the mouth of the container. After the reaction is complete, she again finds the total mass of the system. Which statement is true about this investigation?

A. Mass is gained during the reaction because a chemical change is taking place.

B. The mass does not change because chemical changes observe the law of conservation of mass.

C. Mass is lost during the reaction because the gas is less dense than the baking soda or vinegar.

D. Mass is lost during the reaction because the gas can change in volume but baking soda and vinegar cannot.

Name _____ Date _____

Unit 1

12. Jada and Brian are organizing the information they learned about solids, liquids, and gases. They made a Venn diagram. Now they need to fill it in.

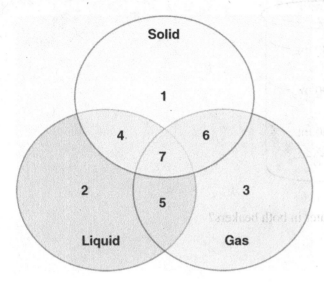

What should Jada and Brian write where all of the states overlap at position 7?

A. Particles are spaced far apart.

B. Particles are constantly moving.

C. Shape depends on the container.

D. Sample does not take shape of container.

13. What happens to the particles of a sample of matter as its temperature increases?

A. The particles melt.

B. The particles move faster.

C. The particles grow in size.

D. The particles become lighter.

14. On a cool morning, a thick fog forms. Through which process does the fog form?

A. Water vapor in the air condenses into liquid drops.

B. Drops of water fall from clouds high up in the atmosphere.

C. Liquid water on the ground evaporates into drops in the air.

D. Air particles change into water molecules as they lose energy.

Name _____ Date _____

Unit 1

15. The two beakers shown below contain pure water.

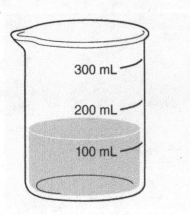 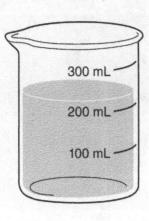

Which of these properties is the same for the water in both beakers?

A. mass

B. weight

C. density

D. volume

Critical Thinking
Answer the following questions in the space provided.

16. A student is investigating the particles in a glass window, raindrops on the window, and the air around the window. Assume the window, the raindrops, and the air are all at 25 °C.

    Identify the state of matter for each sample of matter.

    _____

    _____

    Compare the attractions between the particles in each sample of matter.

    _____

    _____

    _____

    _____

    _____

    _____

Name _____ Date _____

Extended Response
Answer the following questions in the space provided.

17. An ice cube that has a mass of 20 g is in a sealed container. As the container is heated, the ice first melts, but eventually it changes to water vapor.

    The temperature at which the ice melts is _____
    Compare the motion of the particles in each state.
    _____
    _____

    Compare the attraction of the particles in each state.
    _____
    _____

    Compare the mass of the ice cube with the mass of the water vapor.
    _____
    _____

Unit Test B
© Houghton Mifflin Harcourt Publishing Company

Name _____ Date _____

Unit 2

Pretest

Energy

Choose the letter of the best answer.

1. What is an alternative energy resource?

   A. an energy resource used in place of fossil fuels

   B. an energy resource that can be used to make fossil fuels

   C. an energy resource that is used faster than the rate at which it is replaced

   D. an energy resource that can be used without any negative impact on the environment

2. A toy robot can walk and talk and runs on batteries. What type of energy is stored in the batteries?

   A. nuclear

   B. chemical

   C. electrical

   D. mechanical

3. The following diagrams show the particle arrangement in three different states of matter. Which diagram most likely represents the fastest moving particles?

   A.

   B.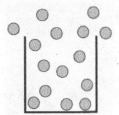

   C.

   D. There isn't enough information to answer this question.

4. The law of conservation of energy describes how the amount of total energy can change in a closed system. Which of these statements correctly describes this law?

   A. The amount of kinetic energy in a system is a constant value.

   B. An object can have potential energy or kinetic energy, but not both at the same time.

   C. Energy can be converted from one form into another, but it cannot be created or destroyed.

   D. The amount of energy created when an object moves is equal to the energy destroyed when it stops moving.

5. Human activities involving fossil fuels can affect the environment in many ways. For example, drilling into land or the ocean floor can destroy habitats and pollute water and soil. Which of the following human activities does this example most likely describe?

   A. using fossil fuels

   B. obtaining fossil fuels

   C. transporting fossil fuels

   D. converting fossil fuels to usable forms

Name _____ Date _____

Unit 2

6. When a ball is thrown into the air, it follows a curved path.

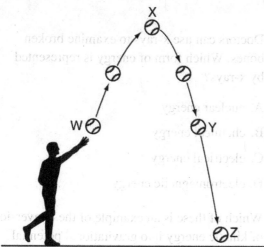

At which point does the ball have the greatest amount of gravitational potential energy?

A. point W
B. point X
C. point Y
D. point Z

7. The freezing point of corn oil is about 259 K. By how many degrees is this different from the freezing point of water on the Kelvin scale?

A. 14 K
B. 47 K
C. 114 K
D. 159 K

8. Which of these choices best describes the thermal energy of a substance?

A. the sum of the kinetic energy of all of the particles of a substance
B. the average kinetic energy of the particles that make up a substance
C. the total kinetic and potential energies of the particles of a substance
D. the difference between the kinetic energy and potential energy of a substance

9. Which change of state takes place when a gas loses energy?

A. melting
B. evaporation
C. solidification
D. condensation

10. As Wilma holds a pan over a fire, the water begins to boil, as shown in the figure below.

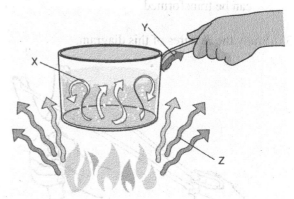

Which process is indicated by the letter X?

A. radiation
B. insulation
C. convection
D. conduction

Pretest
© Houghton Mifflin Harcourt Publishing Company

Module H • Assessment Guide

Name _____ Date _____

Unit 2 Lesson 1
Lesson Quiz

Introduction to Energy

Choose the letter of the best answer.

1. A cold block of metal is placed in contact with a warm block of metal. Which type of energy will be transferred between the blocks?

   A. thermal energy
   B. chemical energy
   C. electrical energy
   D. mechanical energy

2. What is the law of conservation of energy?

   A. The total amount of energy on Earth is constant.
   B. Energy can only be created when a force acts on an object.
   C. Heating an object creates energy and cooling destroys energy.
   D. Energy cannot be created or destroyed but it can be transformed.

3. Study the features of this diagram.

   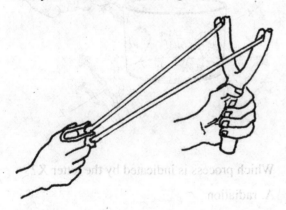

   When the slingshot is held in position as in the diagram, what kind of energy is present?

   A. kinetic energy
   B. potential energy
   C. electrical energy
   D. electromagnetic energy

4. Doctors can use x-rays to examine broken bones. Which form of energy is represented by x-rays?

   A. nuclear energy
   B. chemical energy
   C. electrical energy
   D. electromagnetic energy

5. Which of these is an example of the conversion of kinetic energy into gravitational potential energy?

   A. a bicyclist coasting down a long hill
   B. a car engine moving the car forward
   C. a roller coaster car moving toward the top of a rise in the tracks
   D. a thrown ball falling toward the ground after passing through the top part of its motion

Name _____ Date _____ Date _____

Unit 2 Lesson 2

Lesson Quiz

Temperature

Choose the letter of the best answer.

1. During science class, Sophie measures the temperature of water every minute as it is heating. After a few minutes, the temperature is 82°C. How far below the boiling point of water is this?

    A. 8°C

    B. 18°C

    C. 130°C

    D. 191°C

2. Deval drew the models below of particles in a substance. Which model best represents the particles in a solid?

    A.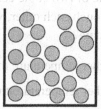

    B.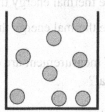

    C.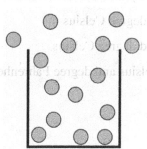

    D.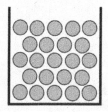

3. Liang is warming a pot of soup on the stove. How is the motion of the particles in the soup different after the temperature of the soup increases?

    A. They move less freely.

    B. They move faster on average.

    C. They have less average energy.

    D. They vibrate and are close together.

4. A scientist warms a chemical to 385 K, and the chemical begins to boil. How much higher is this boiling point of the chemical than the boiling point of water?

    A. 12 K

    B. 73 K

    C. 112 K

    D. 173 K

5. A ball rolls across the grass. Which of these statements best describes what happens to the particles that make up the ball after the ball stops rolling?

    A. The particles are not moving.

    B. The particles are vibrating in place.

    C. The particles are sliding past each other.

    D. The particles are moving from the ball to the grass.

Name _____   Date _____

Unit 2 Lesson 3

Lesson Quiz

Thermal Energy and Heat

Choose the letter of the best answer.

1. Inside a room, the air is often warmer near the ceiling than near the floor. Which of the following accounts for this difference in temperature?

   A. radiation

   B. insulation

   C. convection

   D. conduction

2. Addison warms a pure, solid substance. He records the changes in temperature over time in the left-hand column of the following data table. In the right-hand column, he records the state of the substance.

Temperature (°C)	State
0	solid
X	liquid
Y	boiling liquid
Z	gas

   Which of the following could represent the missing values in Addison's data table?

   A. $X = 50$; $Y = 100$; $Z = 50$

   B. $X = 50$; $Y = 100$; $Z = 150$

   C. $X = -50$; $Y = -100$; $Z = -50$

   D. $X = -50$; $Y = -100$; $Z = -150$

3. How is temperature related to heat?

   A. Temperature is a measure of the heat of an object.

   B. Heat causes a change in the temperature of an object.

   C. Raising the temperature causes the heat of an object to increase.

   D. Temperature and heat are two different ways to measure the same thing.

4. Sarah heated two cubes of aluminum to 50°C. Cube A has a volume of four cubic centimeters. Cube B has a volume of two cubic centimeters. If the cubes do not touch each other, which of these statements is **true**?

   A. Cube A has a higher temperature than cube B.

   B. Cube B has a higher temperature than cube A.

   C. Cube A has more thermal energy than cube B.

   D. Cube B has more thermal energy than cube A.

5. Which two units of measurement are commonly used to measure heat?

   A. calorie and joule

   B. joule and degree Celsius

   C. calorie and degree Celsius

   D. degree Celsius and degree Fahrenheit

Name _____ Date _____

Unit 2 Lesson 4

Lesson Quiz

Effects of Energy Transfer

Choose the letter of the best answer.

1. Through which process do fossil fuels form?

    A. Wind and water gradually break rock into pieces.

    B. Volcanic lava cools and hardens after an eruption.

    C. Heat and pressure act on layers of sediment over time.

    D. Acids in rainwater slowly convert rock to other substances.

2. Which of the following is considered an alternative energy resource?

    A. coal

    B. biomass

    C. petroleum

    D. natural gas

3. Which type of power plant uses one of Earth's nonrenewable resources to produce electrical energy?

    A. wind

    B. solar

    C. nuclear

    D. hydroelectric

4. Where is the energy from the sun stored on Earth?

    A. in rocks

    B. in green plants

    C. in ocean water

    D. in animal bodies

5. Alternative energy sources that are used to generate electricity have both benefits and drawbacks for the environment. The table shows benefits and drawbacks of energy sources used by four communities.

Community	Benefits	Drawbacks
A	is a renewable energy source	can destroy habitats
B	does not release carbon dioxide	produces waste that is difficult to store
C	is a renewable energy source	can only be used near hot springs or volcanoes
D	is a renewable energy source	releases carbon dioxide

    Which community **most likely** uses a hydroelectric power plant to generate most of its electricity?

    A. community A

    B. community B

    C. community C

    D. community D

Name _____ Date _____

Unit 2 Lesson 1

Alternative Assessment

Introduction to Energy

Take Your Pick: *Energy*
Complete the activities to show what you've learned about energy.

1. Work on your own, with a partner, or with a small group.
2. Choose items below for a total of 10 points. Check your choices.
3. Have your teacher approve your plan.
4. Submit or present your results.

2 Points

_____ **Energy Journal** Write a journal entry that describes how energy behaves in a closed system.

_____ **Making a Transformation** Perform a skit in which you identify an instance when one form of energy transforms to another form of energy.

_____ **Energy Observations** Observe your classroom, and note examples of energy. Make a set of diagrams depicting the room. Label and describe at least three examples of energy in the room.

5 Points

_____ **Where Did the Energy Go?** Make sketches that show energy transformations. Make sure you account for all of the initial energy. Think about energy that goes into two different forms and the way people use energy. Label the forms of energy in each sketch.

_____ **Puzzling Terms** Design a crossword puzzle using at least four terms about energy transformation. Be sure to include an answer key with your puzzle.

_____ **Make a Model** Make a pendulum. Use it to explain how mechanical energy works during the pendulum's movement.

8 Points

_____ **Experimenting with Energy** Design an experiment you could conduct that would test the law of conservation of energy. In your design, include the hypothesis you will test and the method you will use to test it. At the beginning of your design, describe the law of conservation of energy.

_____ **Oral Presentation** Prepare an oral presentation to explain how kinetic energy and potential energy differ. Include some of the many forms of kinetic and potential energy in your explanation. Include a demonstration that shows the differences between some of these forms of energy. For example, you might want to use a balloon to illustrate what happens to energy when you fill a balloon with air and then release the balloon.

Name _____ Date _____

Unit 2 Lesson 2
Alternative Assessment

Temperature

Points of View: *Temperature and Measuring Temperature*
Your class will work together to show what you've learned about temperature from several different viewpoints.

1. Work in groups as assigned by your teacher. Each group will be assigned to one or two viewpoints.

2. Complete your assignment, and present your perspective to the class.

 Vocabulary Write and define the vocabulary words from this lesson. Explain what each word has to do with temperature.

 Illustrations Make a poster on which you draw a Fahrenheit, Celsius, and Kelvin thermometer. Color the thermometers so all of them show the temperature at which water freezes. Be sure to label each thermometer.

 Analysis Suppose you are coordinating a community event at a local park. One of your staff members tells you it will be 35 degrees on the day of the event. How should people prepare for the event if it is 35 degrees C, F, or K?

 Observations Observe ice, cool water, and warm water. Use a thermometer to note the temperature of each sample. Write your observations in a notebook. Then apply what you know about kinetic energy to describe how particles behave at each temperature. Use your observations to help explain your ideas.

 Details Create a PowerPoint presentation in which you describe the Fahrenheit, Celsius, and Kelvin temperature scales. In your presentation, give details about freezing and boiling points on each temperature scale.

Name _____ Date _____

Unit 2 Lesson 3

Alternative Assessment

Thermal Energy and Heat

Points of View: *Thermal Energy and Its Transfer*
Your class will work together to show what you've learned about thermal energy transfer from several different viewpoints.

1. Work in groups as assigned by your teacher. Each group will be assigned to one or two viewpoints.

2. Complete your assignment, and present your perspective to the class.

Thermal Energy Transfer

Vocabulary Look up the word *radiation* in the dictionary and write its definition on a sheet of paper. Then look up words that are similar to *radiation*, such as *radiator*, *radiate*, and *radiant*. Write a journal entry in which you compare and contrast the meanings of the related words.

Examples Make a poster that shows items found indoors and out. On your poster, label at least three examples of objects that make good conductors and at least three examples of objects that make good insulators. Describe the qualities that make them good conductors or insulators.

Illustrations Draw a cartoon in which several methods of thermal energy transfer are shown. Include a caption that identifies the methods of transfer and explains how the methods are related.

Analysis Imagine your friend has just ordered hot cocoa at a restaurant. When she tries to take a sip of the drink, she finds it is too hot. Your friend sets her mug on the table. You notice steam rising from the mug. A few minutes later, your friend can drink her cocoa because it has cooled. Use what you know about energy transfer to explain what has happened.

Details Develop a PowerPoint presentation in which you explain how convection works by using ocean water as an example. In your presentation, give details about where the water is moving during convection and why it is moving. Describe how water temperature is measured, and how you would measure to show thermal energy transfer.

Name _____ Date _____

Unit 2 Lesson 4

Alternative Assessment

Effects of Energy Transfer

Tic-Tac-Toe: *Our Energy Use*

1. Work on your own, with a partner, or with a small group.
2. Choose three quick activities from the game. Check the boxes you plan to complete. They must form a straight line in any direction.
3. Have your teacher approve your plan.
4. Do each activity, and turn in your results.

__ **Be the Teacher**	__ **Collage**	__ **Pro/Con Grid**
Write your own quiz to test how well students have learned the material in this lesson. Make sure your questions cover the lesson's key ideas. Include an answer key. Distribute to students and your teacher and see how well they do.	Cut out images related to energy and the environment from magazines and newspapers. Arrange them to make a collage. Include at least six energy resources in your collage and identify them as renewable or nonrenewable.	Create a grid that organizes the disadvantages and advantages for each of the energy resources described in the lesson. Make sure your grid is neat and includes illustrations or diagrams.
__ **Skit**	__ **Commercial**	__ **Poem**
Write and perform a skit in which characters discuss the issues related to using renewable and nonrenewable resources. The topic you cover should include pollution and cost concerns.	Write and perform a commercial for a real or created product that uses an alternative energy source that highlights the benefits of the energy source. Record a video of your commercial if possible.	Choose a poetry form, such as haiku or acrostic, and write a poem about how human energy use can affect the environment.
__ **Invention**	__ **Puzzle Time**	__ **Web Site**
Design a machine that uses one or more alternative energy sources. Sketch a diagram of your invention and include a description of how it would use alternative energy sources and what effect it would have on the environment.	Make a crossword puzzle that uses words, phrases, and ideas from the lesson as solutions for the clues. The puzzle should use at least ten terms, including *renewable*, *nonrenewable*, and *energy*.	Design a web site that focuses on energy use and its effects on the environment. Include a glossary of key terms, images, and examples. Make sure your web site has at least three pages.

Unit 2
Performance-Based Assessment Teacher Notes

Heating Ice Water

Purpose In this activity, students will observe the effects of adding heat energy to a beaker of ice water, and use their knowledge of temperature and changes of state to answer questions.

Time Period 45 minutes

Preparation Equip each activity station with the necessary materials. Position and clip a thermometer into each beaker so that the thermometer is at least 2 cm from the bottom of the beaker.

Safety Tips Remind students to review all safety cautions and icons before beginning this activity. Spilled water is a slipping hazard. Have students inform you of spills immediately. Hot plates are a burn hazard and should be used with care in the presence of water. Have students use caution when using the hot plate, and make sure they are aware of safe practices regarding the use of electrical devices.

Teaching Strategies This activity works best in groups of 2–3 students. The consistent temperature of the ice water demonstrates that temperature does not change until the change of state is complete.

Scoring Rubric

Possible points	Performance indicators
0-10	Appropriate use of materials and equipment
0-50	Quality and clarity of observations
0-40	Analysis of observations

Name _____ Date _____

Unit 2

Performance-Based Assessment

Heating Ice Water

Objective
You have learned about thermal energy, including how adding energy to a substance may cause a change of state. In this activity, you will observe the effects of adding energy to ice water. As you heat the ice water, you will observe and record changes in temperature and state.

Know the Score!
As you work through this activity, keep in mind that you will be earning a grade for the following:

- how well you work with the materials and equipment (10%)
- the quality and clarity of your observations (50%)
- how well you analyze your observations (40%)

Materials and Equipment

- beaker, 250 mL
- hot plate
- ice cubes
- thermometer
- water, cold
- graph paper

Safety Information

- Wipe up spills immediately; spilled water is a slipping hazard.
- Use caution when using the hot plate; the hot plate is a burn hazard.

Procedure

1. What do you predict will happen to the temperature of ice water as the ice melts?

   _____

   _____

2. Fill a beaker about half way with cold tap water. The beaker should already include a thermometer clipped to the side. Make sure the thermometer is at least 2 cm above the bottom of the beaker.

3. Add several ice cubes. Wait until the temperature stabilizes. What is the temperature of the ice water?

   _____

Performance-Based Assessment
© Houghton Mifflin Harcourt Publishing Company

Module H • Assessment Guide

Name _____ Date _____   Unit 2

4. Place the beaker on a hot plate set to low heat. Observe and record the temperature of the water every 15 seconds into the table below. Circle the time at which all of the ice cubes have melted.

Time (min:sec)	Temperature (°C)	Time (min)	Temperature (°C)	Time (min)	Temperature (°C)
0:00		2:00		4:00	
0:15		2:15		4:15	
0:30		2:30		4:30	
0:45		2:45		4:45	
1:00		3:00		5:00	
1:15		3:15		5:15	
1:30		3:30		5:30	
1:45		3:45		5:45	

5. Plot the data on a piece of graph paper. Label the x-axis "Time (min:sec)" and the y-axis "Temperature (°C)". After plotting the data points, draw a trend line.

Analysis

6. Describe the shape of the trend line and explain the relationship between the ice cubes and temperature.

_____

_____

_____

7. Why didn't the temperature of the water change while the ice cubes were melting?

_____

_____

_____

8. Why did the water's temperature rise after the ice cubes melted and the water continued to be heated?

_____

_____

_____

Name _____ Date _____

Unit 2: Energy

Unit 2 — Unit Review

Vocabulary
Check the box to show whether each statement is true or false.

T	F	
☐	☐	1. A <u>fossil fuel</u> is a renewable resource formed from the remains of ancient organisms.
☐	☐	2. <u>Mechanical energy</u> is the sum of an object's kinetic and potential energy.
☐	☐	3. A <u>renewable resource</u> forms at a rate that is much slower than the rate in which the resource is used.
☐	☐	4. The <u>kinetic theory of matter</u> states that all of the particles that make up matter are in a fixed position.
☐	☐	5. <u>Heat</u> is the energy transferred from an object at a higher temperature to an object at a lower temperature.

Key Concepts
Read each question below, and circle the best answer.

6. How could two objects have the same temperature but different thermal energies?

   A. One object could have more heat.

   B. One object could have more calories.

   C. One object could have more particles and lesser total kinetic energy.

   D. One object could have more particles and greater total kinetic energy.

7. What is any energy resource that can be used in place of fossil fuels called?

   A. alternative energy   C. nuclear energy

   B. solar energy   D. biomass energy

8. Energy exists in different forms. Which of the following forms of energy best describes the energy stored in food?

   A. electromagnetic energy   C. sound energy

   B. mechanical energy   D. chemical energy

Name _____ Date _____

Unit 2

9. A mass hanging from a spring moves up and down. The mass stops moving temporarily each time the spring is extended to its fullest at Position 2 and each time it returns to its tight coil at Position 4.

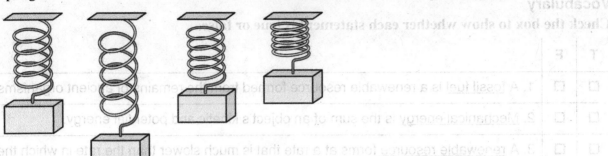

Position 1   Position 2   Position 3   Position 4

Which answer choice best describes the type of energy the spring has at Position 1?

A. potential energy

B. kinetic energy

C. both potential energy and kinetic energy

D. neither potential energy or kinetic energy

10. Which of the following is the transfer of energy as heat by the movement of a liquid or gas?

A. conduction

B. convection

C. emission

D. radiation

11. Which of the following terms means the amount of energy needed to raise the temperature of 1 gram of water by 1 degree Celsius?

A. heat

B. temperature

C. thermal energy

D. calorie

Name _____ Date _____

Unit 2

12. A student collects and records the following data throughout the day.

Time	Temperature (°C)
9 a.m.	12
11 a.m.	14
3 p.m.	16
5 p.m.	13

What instrument did the student use to collect these data?

A. barometer

B. scale

C. thermometer

D. balance

13. What is the difference between a conductor and an insulator?

A. Wood is a good conductor but not a good insulator.

B. Metal is a good insulator but not a good conductor.

C. A conductor transmits energy very well while an insulator does not.

D. An insulator transmits energy very well while a conductor does not.

Critical Thinking

Answer the following questions in the space provided.

14. Describe the law of conservation of energy.

_____

_____

_____

Give two examples of energy being transformed from one type to another.

_____

_____

_____

_____

Name _____ Date _____

Unit 2

15. Three thermometers are lined up side by side.

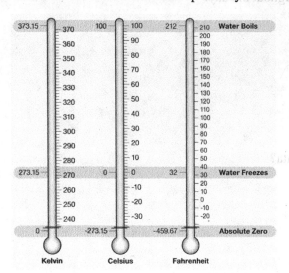

If the temperature outside is 60° F, what is the approximate temperature on the Celsius scale? _____ What is the temperature on the Kelvin scale? _____

If the air temperature drops to 30°F during the night, how has the kinetic energy of the air particles changed?

_____

Connect ESSENTIAL QUESTIONS

Lessons 2 and 3

Answer the following question in the space provided.

16. An ice cube sits in an open container of water placed outside on a sunny day.

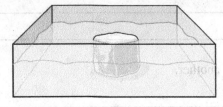

The warmer water contacting the ice cube transfers energy to the ice cube through what process? _____

Use the set-up shown in the diagram to give two examples of how adding energy as heat to a system may result in a change of state.

_____

Compare the speeds of particles in the ice, water, and air.

_____

_____

Unit Review
© Houghton Mifflin Harcourt Publishing Company

Module H • Assessment Guide

Name _____ Date _____

Unit 2

Unit Test A

Energy

Key Concepts
Choose the letter of the best answer.

1. These two beakers contain different amounts of water.

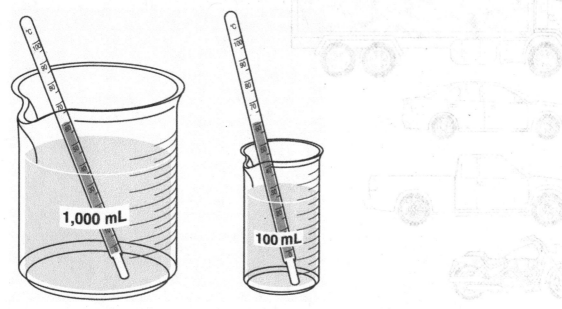

How does the temperature of the water in the large beaker compare or contrast with the temperature of the water in the small beaker?

A. The water in the small beaker has a greater temperature than the water in the large beaker because the water in the small beaker has less mass.

B. The water in the large beaker and the water in the small beaker have the same temperature because the thermometers show the same readings.

C. The water in the large beaker has a greater temperature than the water in the small beaker because the water in the large beaker has more thermal energy.

D. The water in the large beaker has a greater temperature than the water in the small beaker because the water in the large beaker does not cool as quickly when heated.

2. A ball is sitting at the top of a ramp. As the ball rolls down the ramp, the potential energy of the ball decreases. What happens to the potential energy as the ball moves?

A. It is lost as gravitational energy.

B. It is converted to kinetic energy.

C. It is destroyed as the ball moves.

D. It is used to make the ball slow down.

Name _____ Date _____

Unit 2

3. This illustration shows four vehicles. Assume that they are all traveling at the same speed on a highway.

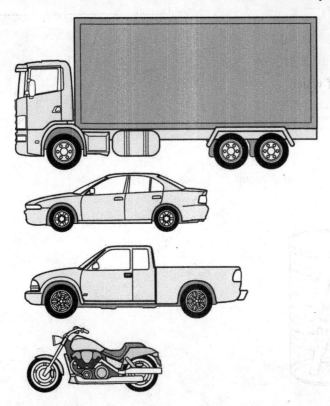

What do you know about the kinetic energy of the vehicles?

A. The motorcycle has the most kinetic energy because it is has the least mass.
B. All of the vehicles have the same kinetic energy because they are moving at the same speed.
C. The delivery van has the greatest kinetic energy because its mass is greater than that of the other vehicles.
D. The delivery van has the greatest kinetic energy because it has the most tires in contact with the pavement.

4. How many calories are needed to raise the temperature of 1 gram of water by 10 degrees Celsius?

A. 0.1 cal
B. 1 cal
C. 10 cal
D. 100 cal

5. Which energy resource supplies most of the energy needs for the United States?

A. biomass
B. fossil fuels
C. solar energy
D. wind energy

Name _____ Date _____

Unit 2

6. As energy in the form of heat is added to an ice cube, it begins to melt. What causes melting?

   A. Heat causes the molecules in the ice cube to expand and forces them apart.

   B. The transfer of thermal energy to the ice cube causes its molecules to move faster.

   C. Heat removes thermal energy from the ice cube and causes it to become liquid water.

   D. The additional energy causes the bonds between hydrogen and oxygen in the ice cube to break.

7. Laith notices that the air in his science classroom is much warmer than the air in his math classroom. Which statement describes how the air particles are different in his colder math classroom?

   A. They are vibrating.

   B. They move more freely.

   C. They move faster on average.

   D. They have less average energy.

8. A student is using the tool below to study a sample of matter.

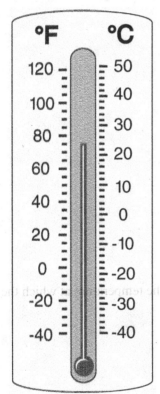

   What property of matter is the student measuring?

   A. mass

   B. density

   C. volume

   D. temperature

Unit Test A
© Houghton Mifflin Harcourt Publishing Company

57

Module H • Assessment Guide

Name _____ Date _____

Unit 2

9. Of the following lists, which is made up of three types of energy?

   A. electrical energy, magnetic energy, sound energy

   B. electronic energy, magnetic energy, thermal energy

   C. geothermic energy, mechanical energy, nuclear energy

   D. electromagnetic energy, mechanical energy, sound energy

10. How is a renewable energy resource different from a nonrenewable energy resource?

    A. Renewable resources come only from plants.

    B. Renewable resources exist in unlimited supplies.

    C. Renewable resources do not have costs associated with them.

    D. Renewable resources can be replaced at the same rate at which they are used.

11. Eduardo puts a beaker of ice and water on a hot plate as shown. After a few minutes, only liquid water is in the beaker. Eventually, the water boils.

    What is the difference between the temperature at which the ice melts and the temperature at which the water boils?

    A. 0 °C

    B. 32 °C

    C. 100 °C

    D. 273 °C

12. How does the use of biomass energy to generate electricity harm the environment the **most**?

    A. Biomass energy production can destroy habitats.

    B. Biomass energy production can create erosion problems.

    C. Biomass energy production can contribute to air pollution.

    D. Biomass energy production can generate hazardous waste.

Name _____ Date _____

Unit 2

Critical Thinking
Answer the following questions in the space provided.

13. Substances freeze and boil at different temperatures.

    What are the freezing and boiling points of water on the Kelvin scale?

    _____

    The temperature of a substance is 75 K. How much higher or lower is this from the freezing point of water and from the boiling point of water?

    _____

    _____

    _____

Extended Response
Answer the following questions in the space provided.

14. A student wants to test heat transfer in different materials. She places four spoons in a bowl of hot water, as shown below. She then makes a data table to record her observations.

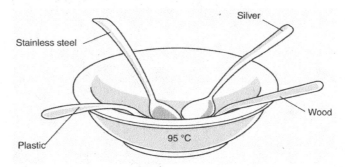

Material	Observation
plastic	slightly warm
stainless steel	hot
silver	very hot
wood	not warm

Identify and describe the type of heat transfer that occurs between the hot water and the spoons.

_____

_____

Unit Test A 59 Module H • Assessment Guide

Name _____ Date _____

Unit 2

After the spoons have been in the water for a couple of minutes, what will the student most likely observe about the temperature of each spoon?

_____

_____

Classify each of the materials that the student is testing as a conductor or an insulator.

_____

_____

_____

_____

Describe how heat transfer due to radiation could be used to affect this system.

_____

_____

Unit Test A
© Houghton Mifflin Harcourt Publishing Company

Module H • Assessment Guide

Name _____ Date _____

Unit 2

Unit Test B

Energy

Key Concepts
Choose the letter of the best answer.

1. These two beakers contain the same liquid substance at the same temperature.

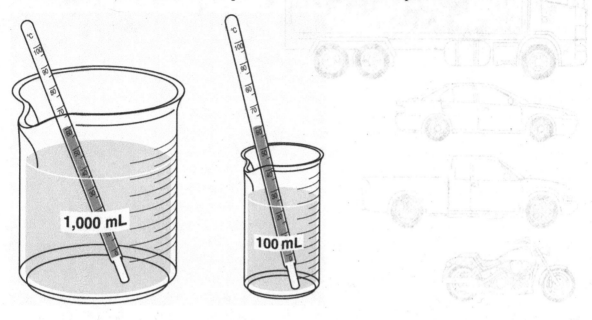

   How does the thermal energy of the liquid in the larger beaker compare to the thermal energy of the liquid in the smaller beaker?

   A. The liquid in the larger beaker has less thermal energy than the liquid in the smaller beaker.

   B. The liquid in the larger beaker has more thermal energy than the liquid in the smaller beaker.

   C. The exact volume of liquid in each beaker must be known to compare the thermal energy of the liquids.

   D. The liquid in the larger beaker has the same amount of thermal energy as the liquid in the smaller beaker.

2. Gordon throws a baseball into the air. It rises, stops when it reaches its greatest height, and then falls back to the ground. At what point does kinetic energy convert to potential energy?

   A. when the baseball is rising

   B. when the baseball is falling

   C. while the baseball sits on the ground

   D. while the baseball is stopped in the air

Name _____ Date _____

Unit 2

3. Assume that all of the vehicles below are traveling on the highway with the same amount of kinetic energy.

Which statement about the vehicles is true?

A. All of the vehicles are traveling at exactly the same speed.

B. The motorcycle is traveling faster than any of the other vehicles.

C. The delivery van is traveling faster than the pickup truck and the car.

D. All four vehicles started traveling at the same time and from the same place.

4. How many calories are needed to raise the temperature of 10 grams of water by 10 degrees Celsius?

A. 1 cal

B. 10 cal

C. 20 cal

D. 100 cal

5. Which of the following is the most common way that obtaining fossil fuels negatively affects the environment?

A. It requires drilling into land or the ocean floor.

B. It disrupts the migration patterns of fish and other animals.

C. It causes the release of greenhouse gases into the atmosphere.

D. It produces radioactive wastes that are harmful for many years.

Name _____ Date _____   Unit 2

6. As liquid water loses energy in the form of heat, the water begins to freeze. What causes freezing?

   A. The loss of thermal energy from the water causes its molecules to move slower.

   B. Heat removes thermal energy from the liquid water and causes it to become an ice cube.

   C. The loss of heat causes the molecules in the ice cube to contract and forces them together.

   D. The lost energy causes the bonds between hydrogen and oxygen in the liquid water to break.

7. While on vacation, Carine rides in a hot-air balloon. To make the balloon rise, the pilot turns on a flame to warm the air inside the balloon. How does the air change?

   A. The temperature of the air decreases.

   B. The air particles move closer together.

   C. The motion of the air particles increases.

   D. The air changes to a different state of matter.

8. The thermometer in the figure below shows the temperature on both the Fahrenheit scale and the Celsius scale.

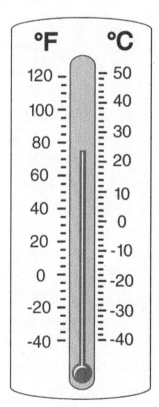

   What temperature does the thermometer show on the Fahrenheit scale?

   A. 70 °F

   B. 75 °F

   C. 80 °F

   D. 85 °F

Name _____ Date _____

Unit 2

9. Which form of energy is due to the motion of an object's particles?

   A. chemical energy

   B. electromagnetic energy

   C. mechanical energy

   D. thermal energy

10. How can a renewable energy resource become nonrenewable?

    A. It is used faster than it can be replaced.

    B. It has negative effects on the environment.

    C. It is purchased and controlled by a single company.

    D. It is found in a location that makes it difficult to obtain.

11. Reggie puts a beaker of ice and water on a hot plate, and the ice slowly begins to melt as shown in figure below. After a few minutes, only liquid water is in the beaker. Eventually, the water becomes hot and boils. The liquid water changes to water vapor, which is a gas.

How is the liquid water, as it is warming, different from the ice?

    A. The freezing point of the ice is lower.

    B. The temperature of the liquid water is lower.

    C. The average kinetic energy of the ice particles is lower.

    D. The motion of the particles that make up the water is slower.

12. Which best describes the use of nuclear energy as an energy source when compared to fossil fuels?

    A. Nuclear energy destroys more habitats.

    B. Nuclear energy produces less air pollution.

    C. Nuclear energy creates more land erosion.

    D. Nuclear energy produces less harmful waste on land.

Unit Test B
© Houghton Mifflin Harcourt Publishing Company

Module H • Assessment Guide

Name _____ Date _____

Unit 2

Critical Thinking
Answer the following questions in the space provided.

13. The temperature of a beaker of water is 87°F.

    How many degrees cooler would the water have to be to freeze?

    _____

    How many degrees warmer would the water have to be to boil?

    _____

Extended Response
Answer the following questions in the space provided.

14. A student in a science lab measures the mass and temperature of water in three different beakers. He records his results in the table shown below.

Beaker	Mass (g)	Temperature (°C)
1	100	70
2	100	50
3	50	50

    Compare the temperatures of the water in each beaker.

    _____

    _____

Name ______________________ Date __________

Compare the average motion of the particles in each beaker.

____

____

Compare the thermal energy of the water in each beaker.

____

____

Compare the flow of energy that will occur if beaker 1 is placed next to beaker 2 with the flow of energy that will occur if beaker 2 is placed next to beaker 3.

____

____

____

____

Name _____ Date _____

Unit 3
Pretest

Atoms and the Periodic Table

Choose the letter of the best answer.

1. The diagram below shows a Bohr model of an atom of nitrogen.

   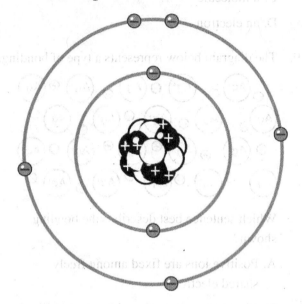

   According to the model, how many valence electrons does the atom have?

   A. 2
   B. 5
   C. 7
   D. 14

2. If you know the atomic number of an element and the element's position in the periodic table, what can you determine about the element?

   A. the atomic mass of the element
   B. the number of neutrons in an atom of the element
   C. the mass number of any element to the left or right of the element
   D. the atomic number of any element to the left or to the right of the element

3. A magnet attracts a piece of metal. How is this event like a chemical bond?

   A. A chemical bond is an interaction that always involves a metal.
   B. A chemical bond is an interaction that holds two atoms together.
   C. In a chemical bond, the substances that come together do not change.
   D. In a chemical bond, the substances that come together can be physically pulled apart.

4. How do most ionic compounds compare with covalent compounds?

   A. Ionic compounds have higher melting and boiling points.
   B. Ionic compounds can generally be reshaped without breaking.
   C. Ionic compounds are more likely to be liquids at room temperature.
   D. Ionic compounds are good insulators of thermal and electrical energy in water.

Pretest 67 Module H • Assessment Guide

Name _____ Date _____

Unit 3

5. Which model of an atom is correctly labeled?

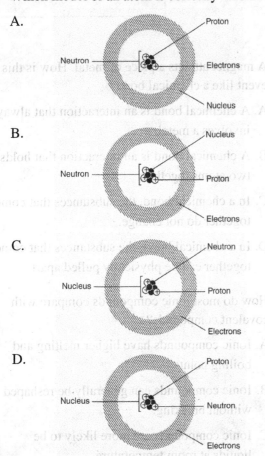

6. What do elements in the same group on the periodic table have in common?

   A. They have the same atomic number.

   B. They have similar chemical symbols.

   C. They have similar chemical properties.

   D. They have the same average atomic mass.

7. Water is made up of units called molecules. Which description best defines a molecule?

   A. an atom that has gained or lost electrons

   B. a solid ionic compound formed from a three-dimensional pattern

   C. an attraction between positively charged metal ions and free electrons

   D. a group of atoms, usually belonging to nonmetals, joined by covalent bonds

8. Which of these is the smallest particle to retain the properties of an element?

   A. an atom

   B. a proton

   C. a molecule

   D. an electron

9. The diagram below represents a type of bonding.

   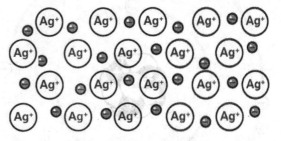

   Which sentence best describes the bonding shown?

   A. Positive ions are fixed among freely shared electrons.

   B. Electrons are transferred from one type of atom to another.

   C. Atoms of different elements share electrons to become stable.

   D. Neutral atoms attract electrons from other elements to become ions.

10. The periodic table of the elements is divided by a line that zigzags along the right side. Which of the following is true for most of the elements that lie along the zigzag line?

    A. They are metals.

    B. They are metalloids.

    C. They are nonmetals.

    D. They are transition metals.

Name _____ Date _____

Unit 3 Lesson 1

Lesson Quiz

The Atom

Choose the letter of the best answer.

1. An atom's mass number is 210 and its atomic number is 85. How many neutrons does the atom have?

   A. 85
   B. 125
   C. 210
   D. 295

2. Think about a wooden chair and a balloon. What do these two objects always have in common?

   A. Both are made of atoms.
   B. Both have the same mass number.
   C. Both are made of the same kind of atom.
   D. Both are made of the same number of atoms.

3. Which of the following scientists added negatively-charged particles to the model of the atom?

   A. Niels Bohr
   B. John Dalton
   C. Ernest Rutherford
   D. J. J. Thomson

4. Calcium is an element. What is the smallest particle of calcium that has all the chemical properties of calcium?

   A. an atom of calcium
   B. a proton from a calcium atom
   C. an electron from a calcium atom
   D. a molecule that contains calcium

5. The diagram below shows a model of an atom.

   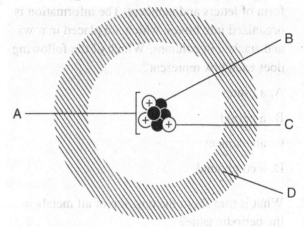

   Which label points to the nucleus?

   A. A
   B. B
   C. C
   D. D

Name ______________________________ Date __________

The Periodic Table

Choose the letter of the best answer.

1. The periodic table has a lot of information in the form of letters and numbers. The information is organized into boxes, which are placed in rows and stacked in columns. Which of the following does each box represent?

   A. a group

   B. a period

   C. an element

   D. a compound

2. What is true about the position of all metals in the periodic table?

   A. They are in the first group.

   B. They are in the bottom period.

   C. They are in the rightmost column.

   D. They are to the left of the zigzag line.

3. The prefix *semi-* means "partly." Based on your knowledge of the properties of elements, which kind of element is most likely used to make semiconductors?

   A. liquids

   B. metals

   C. metalloids

   D. nonmetals

4. In the periodic table below, find the element with the average atomic mass of 87.62.

Part of the Periodic Table

	1	2	3	4	5	6	7	8	9
1	1 **H** 1.008								
2	3 **Li** 6.941	4 **Be** 9.012							
3	11 **Na** 22.99	12 **Mg** 24.31							
4	19 **K** 39.10	20 **Ca** 40.08	21 **Sc** 44.96	22 **Ti** 47.86	23 **V** 50.94	24 **Cr** 52.00	25 **Mn** 54.94	26 **Fe** 55.85	27 **Co** 58.93
5	37 **Rb** 85.47	38 **Sr** 87.62	39 **Y** 88.91	40 **Zr** 91.22	41 **Nb** 92.91	42 **Mo** 95.94	43 **Tc** (98)	44 **Ru** 101.1	45 **Rh** 102.9
6	55 **Cs** 132.9	56 **Ba** 137.3	57 **La** 138.9	72 **Hf** 178.5	73 **Ta** 180.9	74 **W** 183.9	75 **Re** 186.2	76 **Os** 190.2	77 **Ir** 192.2
7	87 **Fr** (223)	88 **Ra** (226)	89 **Ac** (227)	104 **Rf** (263)	105 **Db** (262)	106 **Sg** (266)	107 **Bh** (267)	108 **Hs** (277)	109 **Mt** (268)

Which of the following is the chemical symbol for this element?

   A. Rb

   B. Sc

   C. Sr

   D. Y

5. Which describes how elements are arranged in the periodic table?

   A. by chemical symbol

   B. by increasing atomic mass

   C. by increasing atomic number

   D. by alphabetical order of name

Name _____ Date _____

Unit 3 Lesson 3
Lesson Quiz

Electrons and Chemical Bonding

Choose the letter of the best answer.

1. A student is using colored spheres to represent atoms in a model. In which situation would this type of model be most useful?

   A. representing atoms in a substance in the human body

   B. representing the charged particles that make up individual atoms

   C. representing the individual energy levels in which electrons are found

   D. representing the relationship between electrons and the atomic nucleus

2. Which of the following occurs when substances undergo chemical changes?

   A. Old atoms are destroyed and new atoms are formed.

   B. Atoms change from one element into another element.

   C. Atoms change in size and shape to form new substances.

   D. Bonds between atoms are broken and new bonds are formed.

3. You have been asked to draw a Bohr model of the element carbon. How would you arrange the dots that represent electrons?

   A. They would be embedded in a solid core.

   B. They would be in rings around the nucleus.

   C. They would be located in a central nucleus.

   D. They would be spread evenly in the shape of a cloud.

4. Sodium (Na) is in Group 1 of the periodic table. Sodium atoms readily form bonds with other atoms. Which statement best explains this property?

   A. Sodium boils at a higher temperature than water.

   B. Sodium atoms have an unequal number of protons and neutrons.

   C. Sodium atoms need to give up only one electron to become stable.

   D. Sodium atoms have a lower atomic number than most other elements have.

5. The following image shows a section of the periodic table of the elements.

13	14	15	16	17	18
					2 He
5 B	6 C	7 N	8 O	9 F	10 Ne
13 Al	14 Si	15 P	16 S	17 Cl	18 Ar
31 Ga	32 Ge	33 As	34 Se	35 Br	36 Kr
49 In	50 Sn	51 Sb	52 Te	53 I	54 Xe
81 Tl	82 Pb	83 Bi	84 Po	85 At	86 Rn

   Each atom of sulfur (S) has 6 valence electrons. Which of these elements also has 6 valence electrons?

   A. oxygen (O)

   B. chlorine (Cl)

   C. bromine (Br)

   D. phosphorus (P)

Name _____ Date _____

Unit 3 Lesson 4
Lesson Quiz

Ionic, Covalent, and Metallic Bonding

Choose the letter of the best answer.

1. Which property makes metals good conductors of electricity?

    A. The electrons in a metal can move freely.

    B. The positively charged metal ions attract free electrons around them.

    C. The electrons are shared equally by all of the atoms that make up the metal.

    D. The negative charge of the electrons cancels the positive charge of the metal ions.

2. A student is investigating an ionic compound. Which property should the student expect to observe?

    A. The compound should be a gas at room temperature.

    B. The compound should not dissolve when placed in water.

    C. The compound should have low melting and boiling points.

    D. The compound should conduct electricity when placed in water.

3. Which of the following occurs when a sodium atom forms a positively charged sodium ion?

    A. The atom loses one proton.

    B. The atom gains one proton.

    C. The atom loses one electron.

    D. The atom gains one electron.

4. Between which of these pairs of elements is a covalent bond most likely to form?

    A. carbon and fluorine

    B. sodium and chlorine

    C. potassium and calcium

    D. magnesium and oxygen

5. The diagram below represents carbon dioxide.

    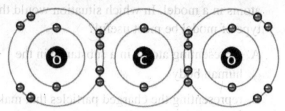

    Based on the type of bonding shown, which term best describes a unit of carbon dioxide?

    A. ion

    B. mixture

    C. element

    D. molecule

Name _____ Date _____

Unit 3 Lesson 1
Alternative Assessment

The Atom

Climb the Pyramid: *Atomic Activities*
Climb the pyramid to show what you have learned about the atom.

1. Work on your own, with a partner, or with a small group.
2. Choose one item from each layer of the pyramid. Check your choices.
3. Have your teacher approve your plan.
4. Submit or present your results.

___ **Atomic Cooking**

Choose an element and write a recipe that describes the parts of its atoms. Your recipe should include the location and charge of protons, neutrons, and electrons, as well as the element's atomic number and mass number.

___ **Atom's Next Top Model**

Construct a 3-D model or mobile for an atom of sodium, oxygen, or magnesium using the element's atomic number and mass number. Your model should show your atom's nucleus and electron cloud, as well as the particles that are located in each part.

___ **Atoms Flipping Out!**

Choose an element and make a flipbook that shows how the electrons move in the electron cloud. Be sure to also identify the nucleus and the particles that can be found inside.

___ **Atoms Acting Out!**

Write and perform a play in which you imagine that you have just developed the atomic theory. Be sure to explain the atomic theory in your play.

___ **Read the Small Print**

Imagine that you are a newspaper writer who is interviewing Democritus or Aristotle about the atom. In your article, discuss how their ideas contributed to the current atomic theory.

___ **That's a Funny Theory**

Design a cartoon or comic strip about the atomic theory. Include some early ideas about the atom, as well as the current atomic theory.

Alternative Assessment
© Houghton Mifflin Harcourt Publishing Company

Module H • Assessment Guide

Name _____ Date _____

Unit 3 Lesson 2

Alternative Assessment

The Periodic Table

Climb the Pyramid: *Elementary Organization*
Climb the pyramid to show what you have learned about the periodic table.

1. Work on your own, with a partner, or with a small group.

2. Choose one item from each layer of the pyramid. Check your choices.

3. Have your teacher approve your plan.

4. Submit or present your results.

___ **Get Organized**

Develop a table that classifies elements as metals, nonmetals, or metalloids. List the name and chemical symbol of each element classified in each group.

___ **All in the Family**

Elements sharing the same group make up a chemical family. Select one group of elements from Groups 1, 2, 13, 14, 15, 16, 17, or 18. Do research to identify six properties shared by the elements making up that group. Construct a display that illustrates the properties of these elements.

___ **Periodic Interview**

Imagine you are a reporter interviewing Dmitri Mendeleev about his contribution to the current arrangement of the elements. Discuss his arrangement, how he came to this conclusion, and whether there are any problems with it. Include an explanation of the current arrangement. Present your newspaper article, T.V. newscast, or radio broadcast to the class.

___ **Body of Elements**

Research what elements make up the human body. Choose three and design a poster in which you show how each is used by the body. Be sure to label the element's name, symbol, atomic number, average atomic mass and in what percentage the element is found.

___ **What's in a Name?**

Choose three elements from the periodic table whose symbols are not made up from letters of the element's name. Research why the chemical symbol for the element was selected. Summarize your findings in a table and include each element's name, symbol, atomic number, and average atomic mass.

___ **Illustrating Mass Differences**

Make a drawing to show the different isotopes of an element that play a role in determining that element's average atomic mass. Label the protons and neutrons in your drawings and identify the atomic number and mass of each element.

Name _____ Date _____

Unit 3 Lesson 3
Alternative Assessment

Electrons and Chemical Bonding

Points of View: *The Interactions of Atoms*
Your class will work together to show what you've learned about how atoms interact from several different viewpoints.

1. Work in groups as assigned by your teacher. Each group will be assigned to one or two viewpoints.

2. Complete your assignment, and present your perspective to the class.

Vocabulary Use the following words to write a story from the point of view of an atom or electron: *chemical bond, chemical change, molecule, valence electrons, periodic table*.

Examples Take several photos inside or outside your school that show chemical changes. Then write or draw descriptions that explain what the chemical change was and how you think it occurred. Include diagrams that show what happened at the molecular level.

Illustrations Draw Bohr models of several atoms from the periodic table. Make sure you put the electrons in the correct energy levels. How many valence electrons does each atom have? What predictions can you make about how likely each atom is to form bonds with other atoms? How can you use the periodic table to determine how the atom will interact with other atoms?

Analysis Think about how electrons move. Can you explain why electrons are sometimes shared while other times they transfer from one atom to another? Share your ideas with the class.

Observations Use a paper cup, a spoonful of baking soda, and vinegar to demonstrate a chemical change. Then research the chemical reaction that takes place. Write a summary of your observations and research. Include atom models of the reactants and products.

Calculations Study some of the elements in Groups 3 through 12. Choose one group and calculate the number of valence electrons for each element in the group. Record your result and note if you observe any patterns.

Name _____ Date _____

Ionic, Covalent, and Metallic Bonding

Unit 3 Lesson 4

Alternative Assessment

Climb the Pyramid: *Bonding*

1. Work on your own, with a partner, or with a small group.

2. Choose one item from each layer of the pyramid. Check your choices.

3. Have your teacher approve your plan.

4. Submit or present your results.

___ **Model**

Make models of an ionic bond, a covalent bond, and a metallic bond. Clearly show atoms, electrons, bonds, and nuclei. Include a key.

___ **Story**

Imagine that you have shrunk, and are small enough to see during ionic, covalent, and metallic bonding. Write a story.

___ **Crossword Puzzle**

Make a crossword puzzle including the lesson's vocabulary terms and five additional related words, such as *atom* and *electron*.

___ **Song**

Write a song about the three types of chemical bonds in a style that you like. Describe each type of bond and its properties.

___ **Flipbook**

Create a flipbook about ionic, covalent, and metallic bonding. Explain how the different bonds form.

___ **Guessing Game**

On index cards, write clues about ionic, covalent, and metallic bonding. Other players must guess the chemical bond the clue is about.

Alternative Assessment
© Houghton Mifflin Harcourt Publishing Company

Module H • Assessment Guide

Using the Periodic Table to Predict Chemical Reactions

Unit 3

Performance-Based Assessment Teacher Notes

Purpose Students will produce and observe a chemical reaction, and use their knowledge of the periodic table and valence electrons to explain the reaction.

Time Period 45-60 minutes

Preparation Equip each activity station with the necessary materials. You may choose to copy the periodic table for students' use during this activity.

Safety Tips Use plastic bowls, if available. Use alcohol thermometers rather than mercury thermometers. Instruct students not to touch broken thermometers. Have a disposal container for sharps available in case of thermometer breakage. Perform this activity in a well-ventilated area. All students should wear safety goggles. Mop up spills immediately.

Teaching Strategies This activity works best in groups of 2–3 students. The reaction of the iron in the steel wool with the oxygen in the air is classified as an oxidation-reduction (redox) reaction. The iron is the reducing agent, which gives up its electrons. The oxygen in the air is the oxidizing agent, which accepts electrons. The acid in vinegar facilitates the transfer of electrons by enhancing the conductivity of the moisture left on the steel wool. You will need to explain to students that the reaction they will observe is between iron and oxygen, not between iron and vinegar. The product of this reaction is iron (III) oxide (rust). Heat is also released during the reaction.

Scoring Rubric

Possible points	Performance indicators
0-30	Appropriate use of materials and equipment
0-40	Quality and clarity of observations
0-30	Explanation of observations

Name _____ Date _____

Unit 3

Performance-Based Assessment

Using the Periodic Table to Predict Chemical Reactions

Objective
In this activity, you will observe a reaction between the iron in steel wool and the oxygen in the air. You will use what you know about valence electrons and the periodic table to predict and explain the chemical reaction you observe.

Know the Score!
As you work through this activity, keep in mind that you will be earning a grade for the following:
- how well you work with the materials and equipment (30%)
- the quality and clarity of your observations (40%)
- how well you use the periodic table to explain those observations (30%)

Materials and Equipment
- bowl, small
- gloves, protective
- graduated cylinder
- rubber band
- steel wool, fine, 0000 grade
- thermometer
- vinegar
- watch or clock

Safety Information
- Be sure to clean up water spills immediately because spilled water is a slipping hazard.
- Do not touch broken thermometers or graduated cylinders.
- Wear safety goggles, aprons, and protective gloves.

Procedure
1. Use what you know about the periodic table and valence electrons to form a hypothesis that answers the question, "What will happen when iron in steel wool reacts with oxygen in the air?"

   _____
   _____

2. Describe the steel wool before the reaction.

   _____
   _____

3. Use the graduated cylinder to measure 50 mL of vinegar. Pour the vinegar into the bowl.

Name _____ Date _____

Unit 3

4. Place the steel wool in the vinegar so that some of the steel wool is touching vinegar and some is not. Leave it there for 2 minutes. Then, with a gloved hand, remove the steel wool from the vinegar.

5. Record the thermometer's starting temperature.

_____

6. Wrap the steel wool around the thermometer. Use a rubber band to hold the steel wool in place.

7. Keep the steel wool around the thermometer for 10 minutes.

8. Record the thermometer's ending temperature.

_____

Analysis

9. Describe the steel wool after the reaction.

_____

_____

10. List the two changes that you observed or measured after the reaction.

_____

_____

11. Use the information about iron and oxygen in the periodic table to identify the number of valence electrons in each element. How does this explain the change in the steel wool's appearance?

_____

_____

12. What would you expect to see if you put steel wool in a chamber filled with chlorine gas instead of oxygen? Explain your answer.

_____

_____

13. What would you expect to see if you put the steel wool in a chamber filled with neon gas? Explain.

_____

_____

14. How does understanding the organization of elements in the periodic table help you make predictions about what is going to happen in an experiment?

_____

_____

Performance-Based Assessment
© Houghton Mifflin Harcourt Publishing Company

Module H • Assessment Guide

Name _____ Date _____

Unit 3: Atoms and the Periodic Table

Unit Review

Vocabulary
Fill in each blank with the term that best completes the following sentences.

1. A(n) _____ is a bond that forms when electrons are transferred from one atom to another.

2. A(n) _____ is an interaction that holds two atoms together.

3. A(n) _____ is the smallest particle of an element that has the chemical properties of that element.

4. A(n) _____ is a bond that forms when atoms share one or more pairs of electrons.

5. A(n) _____ is a negatively-charged subatomic particle.

Key Concepts
Read each question below, and circle the best answer.

6. Which of the following is a property of metals?

    A. low melting point

    B. good electrical conductor

    C. good thermal insulator

    D. cannot bend without breaking

7. The chart below gives the atomic number and mass number of two elements.

	Element A	Element B
Atomic number	10	9
Mass number	20	19

    How many protons does Element B have?

    A. 10   C. 9

    B. 20   D. 19

Name _____ Date _____

8. The diagram below is one model of an atom.

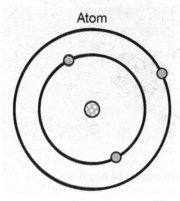

By whom was this model of an atom proposed?

A. Thomson   C. Rutherford

B. Dalton   D. Bohr

9. In a neutral atom, the number of electrons equals the number of what other part?

A. protons   C. nuclei

B. neutrons   D. energy levels

10. Below is a square that represents one element of the periodic table.

```
20
Ca
Calcium
40.078
```

What information is in this square of the periodic table, from top to bottom?

A. average atomic mass, chemical symbol, chemical name, atomic number

B. atomic number, chemical symbol, chemical name, average atomic mass

C. average atomic mass, chemical symbol, chemical name, proton number

D. atomic number, chemical symbol, chemical name, proton number

11. What must happen for an ion to form?

A. An atom must gain or lose an electron.

B. An atom must gain or lose a proton.

C. An atom must gain or lose a neutron.

D. An atom must gain or lose a nucleus.

12. The periodic table is arranged in columns and rows. What are the columns and the rows of the periodic table called?

A. periods and energy levels   C. atomic numbers and periods

B. groups and periods   D. groups and atomic numbers

Name _____ Date _____

Unit 3

13. The diagram below shows how atomic theory has changed over time.

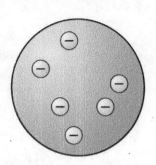 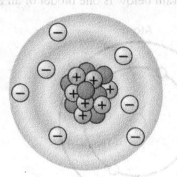

Thomson's model of atom   Rutherford's model of atom   Current model of atom

How is the current understanding of atomic structure different from both Thomson's model and Rutherford's model?

A. Electrons move only in the current model of the atom.

B. Electrons are fixed only in the current model of the atom.

C. Electrons travel in orbits only in the current model of the atom.

D. Electrons move within an electron cloud only in the current model of the atom.

Critical Thinking
Answer the following questions in the space provided.

14. List three properties of metals that nonmetals typically do not have.

_____

_____

Describe where metals and nonmetals are found on the periodic table.

_____

_____

What are elements that have some properties of metals and some properties of nonmetals called?

_____

15. Fluorine has 7 valence electrons. What type of bond is likely to form between two atoms of fluorine?

_____

Draw two atoms of fluorine showing the bond that forms between them.

Unit Review
© Houghton Mifflin Harcourt Publishing Company

82

Module H • Assessment Guide

Name _____ Date _____

Unit 3

How does the number of valence electrons of an atom help to determine whether an atom is likely to form bonds?

_____

_____

Connect ESSENTIAL QUESTIONS

Lessons 1 and 3

Answer the following question in the space provided.

16. In the space below, draw a Bohr model of an atom.

Label the valence electrons. How many valence electrons does this atom have? _____

What element does your atom represent? Explain.

_____

_____

Bohr models do not correctly show the location of electrons in an atom. Explain why they are still useful to predict bonding of atoms.

_____

_____

_____

Name _____ Date _____

Unit 3

Atoms and the Periodic Table

Unit Test A

Key Concepts
Choose the letter of the best answer.

1. Carbon can react with oxygen to form carbon dioxide. Which of the following statements about this chemical change is true?

   A. Carbon and oxygen atoms change into atoms of carbon dioxide.

   B. Carbon and oxygen atoms are rearranged to form a new substance.

   C. Carbon and oxygen atoms are destroyed as new atoms are formed.

   D. Carbon and oxygen atoms have the same properties as molecules of carbon dioxide.

2. An atom of lead has the atomic number 82 and a mass number of 207. How many neutrons does this atom of lead have?

   A. 82

   B. 125

   C. 207

   D. 289

3. The diagram shows a change that occurs to a sodium atom.

   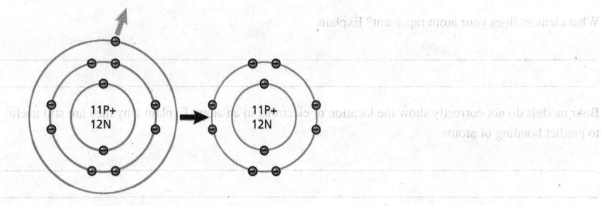

   Sodium

   What is happening in the diagram?

   A. The atom is becoming an ion.

   B. The atom is becoming radioactive.

   C. The atom is changing to a different isotope.

   D. The atom is changing to a different element.

Name _____ Date _____

Unit 3

4. Nitrogen has one more valence electron than carbon has. What can you infer about nitrogen's location on the periodic table?

   A. Nitrogen is in the first group.

   B. Nitrogen and carbon are in different groups.

   C. Nitrogen is directly above carbon but within the same group.

   D. Nitrogen is directly below carbon but within the same group.

5. The segment of the periodic table below shows the elements in Groups 1 through 9.

   Part of the Periodic Table

	1	2	3	4	5	6	7	8	9
1	1 H 1.008								
2	3 Li 6.941	4 Be 9.012							
3	11 Na 22.99	12 Mg 24.31							
4	19 K 39.10	20 Ca 40.08	21 Sc 44.96	22 Ti 47.86	23 V 50.94	24 Cr 52.00	25 Mn 54.94	26 Fe 55.85	27 Co 58.93
5	37 Rb 85.47	38 Sr 87.62	39 Y 88.91	40 Zr 91.22	41 Nb 92.91	42 Mo 95.94	43 Tc (98)	44 Ru 101.1	45 Rh 102.9
6	55 Cs 132.9	56 Ba 137.3	57 La 138.9	72 Hf 178.5	73 Ta 180.9	74 W 183.9	75 Re 186.2	76 Os 190.2	77 Ir 192.2
7	87 Fr (223)	88 Ra (226)	89 Ac<(227)	104 Rf (263)	105 Db (262)	106 Sg (266)	107 Bh (267)	108 Hs (277)	109 Mt (268)

   Which pair of elements are in the same group?

   A. Rb and Cs

   B. Ca and Ti

   C. Li and Be

   D. Na and Mg

6. How do covalent bonds form?

   A. Free electrons move around positive ions.

   B. Outer electrons on some atoms are destroyed.

   C. Electrons are transferred from one atom to another.

   D. Atoms of elements, usually nonmetals, share electrons.

Name _____ Date _____

Unit 3

7. A scientist named J. J. Thomson developed a model of the atom in which electrons are spread throughout a cloud of positive charge. Which discovery directly caused scientists to revise this model?

   A. Electrons were found to have a negative charge.

   B. Atoms for different elements were found to have unique atomic numbers.

   C. A small mass of positive charge was discovered in the center of the atom.

   D. Scientists determined that the exact location of an electron could not be calculated.

8. The diagram shows a portion of the periodic table.

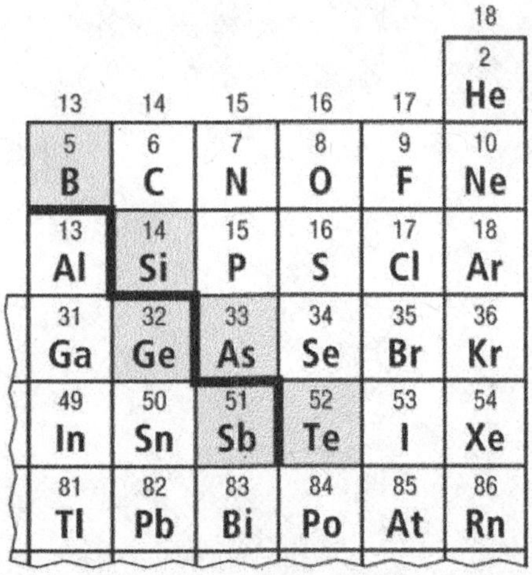

13	14	15	16	17	18
					2 He
5 B	6 C	7 N	8 O	9 F	10 Ne
13 Al	14 Si	15 P	16 S	17 Cl	18 Ar
31 Ga	32 Ge	33 As	34 Se	35 Br	36 Kr
49 In	50 Sn	51 Sb	52 Te	53 I	54 Xe
81 Tl	82 Pb	83 Bi	84 Po	85 At	86 Rn

   What makes the shaded elements unique?

   A. They are metalloids.

   B. They are gases at all temperatures.

   C. They do not react with other elements.

   D. They can be made in a laboratory only.

9. Which of the following is not made up of atoms?

   A. elements

   B. molecules

   C. pure substances

   D. subatomic particles

10. Which properties most likely belong to a metal?

    A. brittle and dull

    B. liquid and clear

    C. ductile and malleable

    D. gaseous and unreactive

Name _____ Date _____

Unit 3

11. How does the strength of a metallic bond compare to other types of bonds?

    A. Metallic bonds are weaker than ionic bonds or covalent bonds.

    B. Metallic bonds are equal in strength to ionic and covalent bonds.

    C. Metallic bonds are weaker than ionic bonds, but stronger than covalent bonds.

    D. Metallic bonds are weaker than covalent bonds, but stronger than ionic bonds.

12. Which of the analogies best describes a Bohr model of an atom?

    A. A Bohr model is like a bowling ball because they are both solid spheres.

    B. A Bohr model is like a model of the solar system because they both show orbits around a massive center.

    C. A Bohr model is like a string of beads because they both contain small parts that are lined up in a row.

    D. A Bohr model is like a jigsaw puzzle because they are both made up of small parts that are all joined together.

Name _____ Date _____

Unit 3

Critical Thinking
Answer the following questions in the space provided.

13. The diagram below shows a portion of the periodic table.

1	
1 H	2
3 Li	4 Be
11 Na	12 Mg
19 K	20 Ca
37 Rb	38 Sr
55 Cs	56 Ba
87 Fr	88 Ra

Based on the periodic table, how many protons does each atom of potassium (K) have? Explain how you know.

_____
_____

How many electrons does each neutral atom of potassium have? Explain how you know.

_____
_____

Name _____ Date _____

Unit 3

Extended Response
Answer the following questions in the space provided.

14. Atoms are made up of smaller particles.

    Identify the three particles within an atom.
    _____
    _____

    Compare these particles in terms of charge.
    _____
    _____
    _____

    Compare these particles in terms of mass.
    _____
    _____
    _____

    Describe how these particles are arranged within the structure of an atom.
    _____
    _____

Name _____ Date _____

Unit 3

Unit Test B

Atoms and the Periodic Table

Key Concepts
Choose the letter of the best answer.

1. When an electric current passes through water (H_2O), it can break apart into hydrogen and oxygen gases. What must be true about this chemical change?

    A. The atoms in water are destroyed and new atoms form.

    B. The gases formed have the same properties as the water.

    C. Water is made up of different atoms than the ones that form hydrogen and oxygen gas.

    D. The number of hydrogen and carbon atoms must be the same before and after the change.

2. What is the mass number of an atom that has 4 protons, 4 electrons, and 5 neutrons?

    A. 4

    B. 5

    C. 9

    D. 13

3. The diagram shows the formation of a sodium ion.

    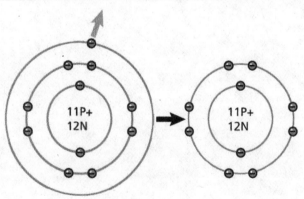

    Sodium

    Which of the following atoms might accept the electron from sodium?

    A. carbon (C)

    B. chlorine (Cl)

    C. copper (Cu)

    D. magnesium (Mg)

Name _____ Date _____

Unit 3

4. Which group of the periodic table has a complete set of valence electrons?

   A. group 1, alkali metals

   B. group 4, carbon group

   C. group 7, halogens

   D. group 8, noble gases

5. The segment of the periodic table below shows the elements in Groups 1 through 9.

   Part of the Periodic Table

	1	2	3	4	5	6	7	8	9
1	1 H 1.008								
2	3 Li 6.941	4 Be 9.012							
3	11 Na 22.99	12 Mg 24.31							
4	19 K 39.10	20 Ca 40.08	21 Sc 44.96	22 Ti 47.86	23 V 50.94	24 Cr 52.00	25 Mn 54.94	26 Fe 55.85	27 Co 58.93
5	37 Rb 85.47	38 Sr 87.62	39 Y 88.91	40 Zr 91.22	41 Nb 92.91	42 Mo 95.94	43 Tc (98)	44 Ru 101.1	45 Rh 102.9
6	55 Cs 132.9	56 Ba 137.3	57 La 138.9	72 Hf 178.5	73 Ta 180.9	74 W 183.9	75 Re 186.2	76 Os 190.2	77 Ir 192.2
7	87 Fr (223)	88 Ra (226)	89 Ac (227)	104 Rf (263)	105 Db (262)	106 Sg (266)	107 Bh (267)	108 Hs (277)	109 Mt (268)

   What do the elements Mg, Ca, and Sr have in common?

   A. The have the same atomic mass.

   B. They have the same mass number.

   C. They have the same atomic number.

   D. They have the same number of valence electrons.

6. Covalent bonds form differently than ionic bonds. How is the difference observed in the properties of covalent substances?

   A. Covalent substances have lower solubility in water.

   B. Covalent substances have higher melting and boiling points.

   C. Covalent substances are brittle and will likely shatter if dropped.

   D. Covalent substances are better conductors of electric current in solution.

Name ______________________________ Date __________

7. The current model of the atom shows electrons in a cloud surrounding the nucleus. How does the electron cloud model differ from the Bohr model?

   A. The electron cloud shows there is not empty space in an atom.

   B. The electron cloud shows that electrons do not have a negative charge.

   C. The electron cloud shows a region in which electrons are likely to be found.

   D. The cloud shows that all the electrons combine together to form one mass.

8. The diagram shows a portion of the periodic table.

13	14	15	16	17	18
					2 He
5 B	6 C	7 N	8 O	9 F	10 Ne
13 Al	14 Si	15 P	16 S	17 Cl	18 Ar
31 Ga	32 Ge	33 As	34 Se	35 Br	36 Kr
49 In	50 Sn	51 Sb	52 Te	53 I	54 Xe
81 Tl	82 Pb	83 Bi	84 Po	85 At	86 Rn

   What property makes some of the shaded elements useful in electronic equipment such as computers?

   A. They are brittle and can break into small pieces easily.

   B. They are gases at room temperature, so they are light in weight.

   C. They have a complete set of valence electrons and do not react.

   D. They can be made to conduct electricity under certain conditions.

9. Atoms are the building blocks of all matter. Which of the following are the building blocks of atoms?

   A. elements

   B. molecules

   C. pure substances

   D. subatomic particles

10. Metals are often used to make pots and pans for cooking. Which property of metals makes them useful for this purpose?

   A. They are shiny.

   B. They are brittle.

   C. They can be drawn into wires.

   D. They are good conductors of thermal energy.

Name _____ Date _____

Unit 3

11. Metals contain free electrons. Why aren't metals negatively charged as a result?

   A. The electrons easily flow out of the metal onto other surfaces.

   B. Metallic bonds pull the charge away from the electrons and toward the metal ions.

   C. The negative charge of the electrons is balanced by the positive charge of the metal ions.

   D. Electrons lose their negative charge as soon as they are separated from their metal atoms.

12. How are Bohr models of atoms useful if they do not show the true arrangement of particles?

   A. They are useful for predicting how atoms will bond.

   B. They are useful for measuring the masses of electrons.

   C. They are useful for showing how energy is stored in the nuclei of atoms.

   D. They are useful for identifying the specific places in which electrons are located.

Name _____ Date _____

Unit 3

Critical Thinking
Answer the following questions in the space provided.

13. The diagram below shows a portion of the periodic table.

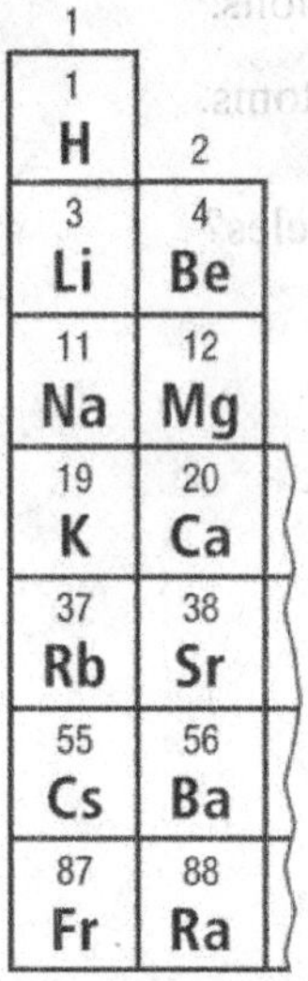

1	
1 H	2
3 Li	4 Be
11 Na	12 Mg
19 K	20 Ca
37 Rb	38 Sr
55 Cs	56 Ba
87 Fr	88 Ra

Based on the periodic table, how is the number of valence electrons different for potassium (K) than it is for calcium (Ca)? Explain how you know.

_____
_____

Use the numbers of valence electrons to explain how atoms of each element could form bonds with other atoms.

_____
_____
_____

Name _____ Date _____ Date _____

Unit 3

Extended Response
Answer the following questions in the space provided.

14. Atoms are made up of smaller particles.

    How does the structure of an atom relate to its atomic number?

    _____
    _____

    How does the structure of an atom relate to its mass number?

    _____
    _____

    Where is most of the mass of an atom concentrated? Explain.

    _____
    _____

    Why are atoms electrically neutral if they contain charged particles?

    _____
    _____

Unit Test B 95 Module H • Assessment Guide

Name _____ Date _____

Unit 4

Interactions of Matter

Pretest

Choose the letter of the best answer.

1. The diagram shows reactants in the presence of a catalyst.

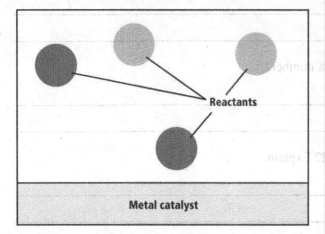

Which is true of this catalyst as the reaction proceeds?

A. It can decrease the rate of reaction by lowering the temperature of the reaction.

B. It can decrease the rate of reaction by decreasing the surface area of the reactant.

C. It can increase the rate of a reaction by bringing reactant particles together at its surface.

D. It can increase the rate of reaction by reacting with other reactants form different products.

2. Solid magnesium, Mg, reacts with oxygen gas, O_2, to form the solid magnesium oxide, MgO. Which equation is a balanced chemical equation for this chemical change?

A. Mg + O --> MgO

B. Mg + O_2 --> MgO_2

C. 2Mg + O_2 --> 2MgO

D. Mg + 2O --> 2MgO

3. What does an alpha particle consist of?

A. two electrons

B. two protons and two neutrons

C. one positron

D. two neutrons and two electrons

4. The diagram shows the three isotopes of hydrogen.

Three Isotopes of Hydrogen

How are the isotopes different?

A. Each isotope has a different number of nuclei.

B. Each isotope has a different number of protons.

C. Each isotope has a different number of neutrons.

D. Each isotope has a different number of electrons.

5. Which formula represents a hydrocarbon?

A. Butane (C_4H_{10})

B. Methanal (HCHO)

C. Ethanal (CH_3CHO)

D. Propanoic acid (CH_3CH_2COOH)

6. What is the difference between an exothermic reaction and an endothermic reaction?

   A. Endothermic reactions release energy; exothermic reactions absorb energy.

   B. Exothermic reactions release energy; endothermic reactions absorb energy.

   C. Endothermic reactions create energy; exothermic reactions destroy energy.

   D. Exothermic reactions create energy; endothermic reactions destroy energy.

7. A student observed a precipitate form during a chemical reaction. Which of the following did the student most likely observe?

   A. Bubbles of gas floated to the top of a solution.

   B. The color of a liquid changed when a solid was added to it.

   C. Steam rose from a flask when a powder was added to a liquid.

   D. A solid sank to the bottom of a container when two liquids were mixed.

8. In a nuclear reactor, the total mass of the particles produced by nuclear fission is less than the total mass of the particles before the reaction. Why?

   A. A small amount of mass is converted to energy.

   B. A small amount of mass escapes the nuclear reactor as a gas.

   C. Larger particles are squeezed to become smaller particles of lesser mass.

   D. The tools needed to measure the mass difference are not precise enough.

9. The diagram shows several types of bonds that carbon can form.

   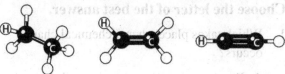

   Single bonding   Double bonding   Triple bonding

   Which conclusion can you reach based on the diagram?

   A. A carbon atom can make up to eight bonds with other atoms.

   B. Carbon atoms increase the number of electrons they can share as the number of bonds increases.

   C. As the number of bonds between carbon atoms increases, the size of the hydrogen atoms decreases.

   D. As the number of bonds between carbon atoms increases, the number of hydrogen atoms in the molecule decreases.

10. How is mass converted to energy during a nuclear fusion reaction?

    A. The nucleus of a large atom splits into two smaller nuclei.

    B. An atom becomes less massive by converting its electrons into energy.

    C. An atom becomes more massive by attracting electrons from other atoms.

    D. The nuclei of smaller atoms combine to form a new, more massive nucleus.

Name _____ Date _____

Unit 4 Lesson 1
Lesson Quiz

Chemical Reactions

Choose the letter of the best answer.

1. Which takes place when a chemical change occurs?

   A. Bonds connecting atoms in products and reactants form.

   B. Bonds connecting atoms in products and reactants break.

   C. Bonds connecting atoms in reactants break, and bonds connecting atoms in products form.

   D. Bonds connecting atoms in reactants form, and bonds connecting atoms in products break.

2. In photosynthesis, water (H_2O) and carbon dioxide (CO_2) react to form oxygen (O_2) and glucose ($C_6H_{12}O_6$). The following table shows the reactants and products of photosynthesis.

Reactants	Products
$6\ H_2O + 6\ CO_2$	$6\ O_2 + ?\ C_6H_{12}O_6$

   What number should replace the question mark to show the number of $C_6H_{12}O_6$ (glucose) molecules in a balanced chemical equation?

   A. 1

   B. 2

   C. 6

   D. 12

3. Which of the following is an example of an endothermic reaction?

   A. a cake baking

   B. a candle burning

   C. a firework exploding

   D. a piece of wood smoldering

4. Why does increasing temperature generally increase the rate of a chemical reaction?

   A. It increases the amount of reactant.

   B. It decreases the size of the reactant particles.

   C. It causes the particles of the reactants to move faster and collide more often.

   D. It increases the space between reactant particles so they have more room to move apart.

5. When coal is burned, sulfur is released into the air, where it reacts with oxygen to produce sulfur dioxide and sulfur trioxide. Which of the following is true of this process?

   A. The identity of each reactant remains the same.

   B. Sulfur dioxide and sulfur trioxide are the reactants in this change.

   C. The products have different chemical properties than those of the reactants.

   D. The products contain more sulfur atoms and fewer oxygen atoms than the reactants.

Name _____ Date _____

Unit 4 Lesson 2

Lesson Quiz

Organic Chemistry

Choose the letter of the best answer.

1. The diagram shows the structural formula for propane.

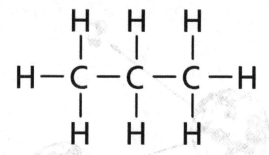

   Based on the diagram, which of the following is the molecular formula for propane?

   A. CH

   B. C_3H

   C. C_3H_3

   D. C_3H_8

2. Proteins, such as hair and nails, are polymers. Which of the following models best represents a polymer?

   A. a chain made of identical links

   B. a car made of many different parts

   C. a bowling ball made of a solid material

   D. a carpet made of many individual fibers

3. Which term best describes a carbon atom with an oxygen atom and an –OH bonded to it?

   A. a polymer

   B. an alcohol

   C. a carboxyl group

   D. an aromatic compound

4. Which of the following elements is carbon most likely to bond with to form organic compounds?

   A. silver

   B. helium

   C. sodium

   D. hydrogen

5. What unique property of carbon enables it to form a huge variety of compounds?

   A. Carbon is a solid at room temperature.

   B. Carbon is a nonmetal that is dull and brittle.

   C. Carbon atoms can have different numbers of neutrons.

   D. Carbon atoms can form four bonds with up to four other atoms.

Name _____ Date _____

Unit 4 Lesson 3

Lesson Quiz

Nuclear Reactions

Choose the letter of the best answer.

1. Which would be a benefit of using nuclear fusion over nuclear fission to generate electricity?

   A. Sustaining fusion reactions is easier and less expensive.

   B. The end products of fusion reactions are not radioactive.

   C. The energy produced by fusion does not need to be contained.

   D. Fusion reactions require conditions of extreme pressure and temperature.

2. How is a nuclear reaction different from a chemical reaction?

   A. Nuclear reactions can release energy.

   B. Nuclear reactions do not release energy.

   C. Nuclear reactions can change the identity of atoms.

   D. Nuclear reactions occur only in nuclear power plants.

3. Which process do nuclear power plants use to produce electricity?

   A. beta decay

   B. gamma decay

   C. fusion reactions

   D. fission reactions

4. The diagram shows a radioactive nucleus undergoing alpha decay.

   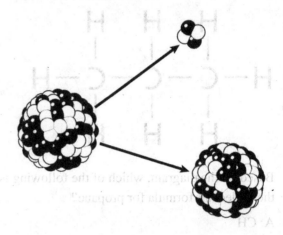

   Which of the following is not given off during this reaction?

   A. energy

   B. an alpha particle

   C. a smaller nucleus

   D. a gamma ray

5. How are isotopes of the same element different from one another?

   A. They have different atomic numbers.

   B. They have different numbers of neutrons.

   C. They have different overall electric charges.

   D. They have different numbers of electrons.

Name _____ Date _____ Date _____

Unit 4 Lesson 1

Alternative Assessment

Chemical Reactions

Points of View: *Looking at Chemical Reactions*
Your class will work together to show what you've learned about chemical reactions from several different viewpoints.

1. Work in groups as assigned by your teacher. Each group will be assigned to one or two viewpoints.

2. Complete your assignment, and present your perspective to the class.

Vocabulary Write a sentence for each of the following vocabulary terms: *chemical reaction, chemical formula, chemical equation, reactant, product, law of conservation of mass, endothermic reaction, exothermic reaction, law of conservation of energy.*

Examples Take several pictures inside or outside your school that show evidence of a chemical reaction. Then write descriptions or draw diagrams to explain what the chemical reaction was and whether the reaction was endothermic or exothermic.

Analysis Think about the energy that is used to move people in cars, trains, and buses. What is the source of this energy? Is the law of conservation of energy observed? Share your ideas with the class.

Observations Dip a steel wool pad into a solution of vinegar and water. Then, set the steel wool in a paper cup and observe several times over the next two hours and again the next morning. Did a chemical reaction take place? Provide evidence that supports your claim.

Calculations List some chemical reactions. Choose one and write the reaction in the form of a chemical equation. Make sure the equation is balanced.

Name _____ Date _____

Unit 4 Lesson 2

Alternative Assessment

Organic Chemistry

Take Your Pick: *Carbon Bonding*
Without the element carbon, life on Earth would be very different.

1. Work on your own, with a partner, or with a small group.
2. Choose items below for a total of 10 points. Check your choices.
3. Have your teacher approve your plan.
4. Submit or present your results.

2 Points

_____ **In Your Words** In your own words, write a definition for the term *organic compound*.

_____ **Why Carbon?** What makes carbon so important to the chemistry of living organisms?

5 Points

_____ **Paper Chains** Cut strips of blue, yellow, and red paper. Staple the strips into paper chains to create models of three types of organic molecules. Blue strips represent carbon, yellow strips represent hydrogen, red strips represent oxygen, and the staples represent the bonds. Label each molecule.

_____ **Gumdrop Molecules** Gather blue, red, and green gumdrops and some toothpicks. Use the gumdrops to represent atoms and the toothpicks to represent bonds. Make a model of a single, a double, and a triple carbon bond. Write the chemical formula for each molecule.

_____ **Diagramming Bonds** Use markers to diagram the bonds in carbon dioxide (CO_2) and ethane (C_2H_6).

_____ **What and Where** List the six elements that make up most molecules in living things. Then tell where in the human body each element is found.

8 Points

_____ **Writing from Research** Pick one important organic compound, research it, and write a report on it. Include in your paper the history of its discovery, its function in organisms, and its use and manufacture, if applicable. Include a diagram of the compound in your paper.

_____ **Polymer Posters** Research one type of polymer such as polyvinyl chloride (PVC), Kevlar, or nylon. Find out how this substance is being used to create new products. Make a poster that tells about the polymer's function in a product.

Name _____ Date _____

Unit 4 Lesson 3

Alternative Assessment

Nuclear Reactions

Climb the Pyramid: *Changing the Nucleus*
Select options at each level to show what you know about nuclear reactions.

1. Work on your own, with a partner, or with a small group.
2. Choose one item from each layer of the pyramid. Check your choices.
3. Have your teacher approve your plan.
4. Submit or present your results.

__ Illustrating Isotopes

Use the periodic table to find the atomic number of carbon. Then create a labeled diagram to show how the isotopes carbon-12 and carbon-14 differ from one another.

__ Making Models

Using any materials you think are appropriate, make models to represent alpha decay, beta decay, and gamma decay. Your models should clearly show the differences among the three types of nuclear decay.

__ Decay Poster

Make a poster that compares alpha particles, beta particles, and gamma rays. Add labels and captions as needed to clearly show the differences among the three types of nuclear radiation.

__ Model It!

Develop a three-dimensional model that can be used to illustrate nuclear fission of a uranium-235 nucleus. Choose materials that are suitable to represent the protons and neutrons involved in the reaction.

__ Concept Mapping

Create a concept map to explain nuclear fission and nuclear fusion. Use labels to describe what happens in each type of reaction.

__ Radioactivity

Make an illustrated poster or other visual display to show how radioactivity is used in the home, in industry, in medicine, and as a source of energy. Be sure to include at least one use from each category.

Alternative Assessment
© Houghton Mifflin Harcourt Publishing Company

Module H • Assessment Guide

Is It a Chemical Reaction?

Unit 4

Performance-Based Assessment Teacher Notes

Purpose Students mix solutions containing unknown chemicals and observe the results. Students determine whether or not a reaction has taken place based on observation.

Time Period 45-60 minutes

Preparation Equip each activity station with five 100–200 mL beakers and "unknown" bottles marked as follows:

- Liquid A: containing 0.5 mL 0.1 M hydrochloric acid
- Liquid B: containing 0.5 mL 0.1 M sodium hydroxide
- Liquid C: containing 0.5 mL 0.1 M silver nitrate
- Liquid D: containing a dilute solution of phenolphthalein (solid phenolphthalein should be dissolved in alcohol, then diluted in water)
- Solid E: containing small pieces of solid calcium carbonate or chalk

Safety Tips Students should not allow unknown solutions to come in contact with their skin or clothing. Students should wear safety goggles whenever they work with chemicals. Instruct students never to smell unknown chemicals or bring them close to their faces. Students should remove jewelry, tie back long hair, and tie down any loose clothing. Provide protective gloves and lab aprons.

Teaching Strategies This activity works best in groups of 4–5 students. The reaction of hydrochloric acid, HCl, and silver nitrate, $AgNO_3$, will produce a solid precipitate of silver chloride, AgCl. The reaction of HCl with calcium carbonate, $CaCO_3$, will produce carbonic acid, H_2CO_3, which decomposes to CO_2 gas. Sodium hydroxide causes the pH indicator phenolphthalein to turn pink.

Scoring Rubric

Possible points	Performance indicators
0-30	Appropriate use of materials and equipment
0-30	Quality and clarity of observations
0-40	Explanation of observations

Name _____   Date _____

Unit 4
Performance-Based Assessment

Is It a Chemical Reaction?

Objective
You have learned that during a chemical reaction, atoms are rearranged in different ways to form new substances with unique physical and chemical properties. Clues that a chemical reaction has taken place include a color change, precipitate formation, an odor, or formation of gas bubbles. In this activity, you will observe chemical reactions and use these clues to decide if a chemical reaction has taken place.

Know the Score!
As you work through this activity, keep in mind that you will be earning a grade for the following:
- how well you work with materials and equipment (30%)
- how well you state your observations (30%)
- how well you use the periodic table to explain those observations (40%)

Materials and Equipment
- beakers, 250 mL (5)
- chemicals, unknown
- droppers (4)

Safety Information
- Do not allow unknown solutions to come in contact with your skin or clothing!
- Never smell unknown chemicals or bring them close to your face!
- Remove jewelry, tie back your hair, and tie down any loose clothing.

Procedure
1. Your station has 5 beakers. Number the beakers 1–5, and place a few drops of Liquid A into beakers 1–4.

2. Add a few drops of Liquid B to the liquid in beaker 1. What do you observe?

   _____

3. Add a few drops of Liquid C to the liquid in beaker 2. What do you observe?

   _____
   _____

4. Add a few drops of Liquid D to the liquid in beaker 3. What do you observe?

   _____

Performance-Based Assessment
© Houghton Mifflin Harcourt Publishing Company

Module H • Assessment Guide

Name _____ Date _____

Unit 4

5. Add some of Solid E to the liquid in beaker 4. What do you observe?
   _____

6. Put a few drops of Liquid B in beaker 5. Add a small quantity of Liquid D. What happens?
   _____

Analysis

7. Transfer your observations in steps 2–6 to the table below.

Chemical Reactions

	Liquid B	Liquid C	Liquid D	Solid E
Liquid A				
Liquid B				

8. When you mixed Liquid A and Liquid B, did a reaction take place? How do you know?
   _____
   _____

9. When you mixed Liquid A and Liquid C, did a reaction take place? How do you know?
   _____
   _____
   _____

10. What happens in the reaction between Liquid A and Solid E that causes the effect you observed?
    _____
    _____

11. How do the properties of the new substance resulting from a chemical reaction compare to those of the original substances in the reaction?
    _____
    _____
    _____

Performance-Based Assessment
© Houghton Mifflin Harcourt Publishing Company

Module H • Assessment Guide

Name _____ Date _____

Unit 4: Interactions of Matter

Unit Review

Vocabulary
Check the box to show whether each statement is true or false.

T	F	
☐	☐	1. In a <u>chemical reaction,</u> atoms are rearranged and bonds can be broken or formed.
☐	☐	2. <u>Nuclear fission</u> is the process by which the nuclei of smaller atoms combine to form a new, more massive nucleus.
☐	☐	3. A <u>nuclear reaction</u> changes the number of electrons in the nucleus.
☐	☐	4. An <u>endothermic reaction</u> releases energy.
☐	☐	5. A <u>hydrocarbon</u> is any substance that contains oxygen and carbon.

Key Concepts
Read each question below, and circle the best answer.

6. What is an atom that has the same number of protons as other atoms of the same element but a different number of neutrons?

   A. catalyst   C. reactant

   B. isotope   D. product

7. Which of the following can speed up the rate of a chemical reaction?

   A. removing a catalyst

   B. lowering the reactant concentration

   C. lowering the temperature

   D. breaking up a reactant into smaller pieces

8. What is a limitation to using nuclear fusion for energy?

   A. Nuclear fusion produces a large amount of radioactive waste.

   B. The hydrogen fuel needed for the reaction is hard to collect.

   C. The atoms needed in nuclear fusion reactions are too small to use.

   D. Creating the proper pressure and temperature conditions on Earth is difficult.

Name _____ Date _____

Unit 4

9. The chemical formula of acetate is shown below.

$$C_2H_3O_2$$

How many atoms of carbon are in one molecule of acetate?

A. 1
B. 2
C. 3
D. 7

10. Carbon, hydrogen, and oxygen are always found in what type of molecule?

A. hydrocarbon
B. carbohydrate
C. acid
D. polymer

11. A chemical reaction is shown below.

$$Fe + H_2O \rightarrow Fe_3O_4 + H_2$$

What are the products in the equation?

A. Fe_3O_4 only
B. Fe only
C. Fe and H_2O
D. Fe_3O_4 and H_2

12. Which of the following occurrences indicates that a chemical reaction has taken place?

A. An odor is produced by burning a sugar cube.
B. A puddle is produced by melting an ice cube.
C. A loud noise is produced by crushing a can.
D. A shard of glass is produced by breaking a bottle.

13. Which of the following is an example of an exothermic chemical reaction?

A. photosynthesis
B. burning wood
C. melting ice cubes
D. boiling water

Unit Review
© Houghton Mifflin Harcourt Publishing Company

Module H • Assessment Guide

Name _____ Date _____

Unit 4

Critical Thinking
Answer the following questions in the space provided.

14. The diagram below shows a nuclear reaction.

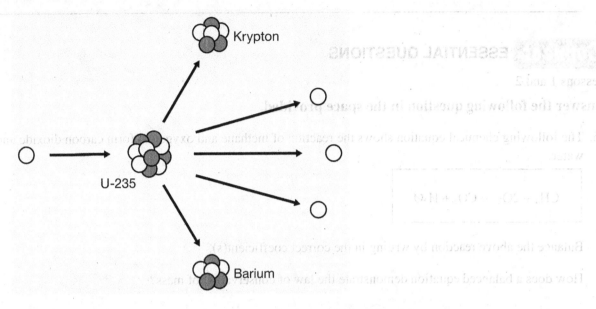

Does the diagram show an example of nuclear fission or fusion? Explain.

_____
_____

Using the diagram, explain how a nuclear chain reaction could occur.

_____
_____
_____
_____

15. What are the three types of radioactive decay?

_____
_____

What is one way in which radioactive decay is used to benefit society?

_____
_____
_____
_____

Unit Review
© Houghton Mifflin Harcourt Publishing Company

Module H • Assessment Guide

Name _____ Date _____ Date _____

Unit 4

In your opinion, do the risks of using radioactive decay outweigh the benefits? Explain.

_____

_____

_____

Connect ESSENTIAL QUESTIONS

Lessons 1 and 2

Answer the following question in the space provided.

16. The following chemical equation shows the reaction of methane and oxygen to form carbon dioxide and water.

    $$CH_4 + 2O_2 \rightarrow CO_2 + H_2O$$

    Balance the above reaction by writing in the correct coefficient(s).

    How does a balanced equation demonstrate the law of conservation of mass?

    _____

    _____

    What molecule in the above equation is a hydrocarbon?

    _____

    Draw a full structural formula for the hydrocarbon that you listed above.

Name _____ Date _____

Interactions of Matter

Unit 4

Unit Test A

Key Concepts
Choose the letter of the best answer.

1. Which of the following foods contains complex carbohydrates?

    A. potato

    B. fruit juice

    C. corn syrup

    D. table sugar

2. Ethane (C_2H_6) is one of many compounds that burns in oxygen to form carbon dioxide and water, as represented by the equation below. Which substance is a product in this reaction?

    $$C_2H_6 + O_2 \rightarrow CO_2 + H_2O$$

    A. carbon (C)

    B. oxygen (O_2)

    C. water (H_2O)

    D. ethane (C_2H_6)

3. Which of the following best distinguishes a nuclear reaction from other types of reactions?

    A. a reaction in which energy is released

    B. a reaction in which electrons are shared

    C. a reaction in which electrons are transferred

    D. a reaction that affects the nucleus of an atom

Name _____ Date _____

4. The following image shows a chemical reaction taking place.

What can you infer about this reaction?

A. Mass is not conserved.

B. The reaction is exothermic.

C. The reaction is endothermic.

D. There are reactants but no products.

5. Which change will most likely lead to an increase in the rate of a chemical reaction?

A. removing a catalyst

B. lowering the temperature

C. increasing the surface area of reactants

D. decreasing the concentration of reactants

6. Glucose ($C_6H_{12}O_6$) is a kind of sugar molecule. How many atoms are in a molecule of glucose?

A. 6

B. 12

C. 18

D. 24

Name _____ Date _____

Unit 4

7. The following chemical equations represent two different chemical reactions.

   > $3NaOH + H_3PO_4 \rightarrow Na_3PO_4 + 3H_2O$
   >
   > $2Mg + O_2 \rightarrow 2MgO$

   Based on these equations, which statement is true?

   A. MgO is a reactant in one of the chemical reactions.

   B. Na_3PO_4 is a product in one of the chemical reactions.

   C. H_3PO_4 shows that three atoms of oxygen are involved in the reaction.

   D. O_2 is a catalyst.

8. Which statement about carbon molecules is true?

   A. They may take the form of rings.

   B. They never form branched chains.

   C. They always take the form of long chains.

   D. They rarely contain more than two carbon atoms.

9. Carbon-6 undergoes beta-minus decay, during which a neutron becomes a proton, an electron, and an antineutrino. What happens to the carbon atom during this process?

   A. Its mass number increases by 1.

   B. Its mass number decreases by 1.

   C. Its atomic number increases by 1.

   D. Its atomic number decreases by 1.

10. What is the source of the large amount of energy produced in a nuclear fission reaction?

    A. Some mass is converted to energy.

    B. Energy for fission reactions is provided by the sun.

    C. The Earth's oceans provide water used in fission reactions.

    D. Electrons from large atoms are converted into gamma rays.

11. The diagram shows the structural formula for ethane.

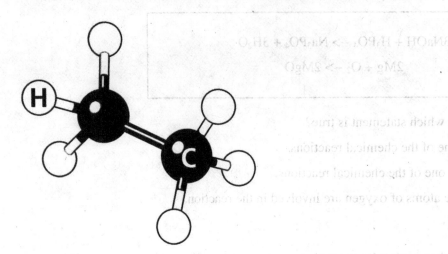

Based on the diagram, which of the following is the chemical formula for ethane?

A. CH_3

B. C_2H_6

C. C_2H_2

D. C_6H_6

12. What causes an atom to undergo radioactive decay?

A. An unstable atom decays to become stable.

B. An isotope decays to form a different type of isotope.

C. An atom with a small nucleus decays to form a larger nucleus.

D. An atom with a low number of electrons decays to gain electrons.

Name _____ Date _____

Unit 4

Critical Thinking
Answer the following questions in the space provided.

13. Many common materials are made up of polymers.

    What is a polymer?
    _____
    _____

    What are the structural units of polymers called?
    _____

    What are two examples of polymers?
    _____
    _____

Name _____ Date _____

Unit 4

Extended Response
Answer the following questions in the space provided.

14. The diagram shows a helium nucleus that formed by nuclear fusion.

What are the mass number and charge of this nucleus?
_____

Explain what happens during nuclear fusion.
_____
_____

How is mass related to energy in this reaction?
_____
_____
_____

Name _____ Date _____

Interactions of Matter

Unit 4

Unit Test B

Key Concepts
Choose the letter of the best answer.

1. How are complex carbohydrates different from simple sugars?

    A. Complex carbohydrates are made up of larger atoms.

    B. Complex carbohydrates are polymers of simple sugars.

    C. Complex carbohydrates have a different ratio of atoms.

    D. Complex carbohydrates are made up of different elements.

2. Ethane (C_2H_6) is one of many compounds that burns in oxygen to form carbon dioxide and water. What is true of the chemical equation shown below?

    $$C_2H_6 + O_2 \rightarrow CO_2 + H_2O$$

    A. The positions of the reactants and the products are reversed.

    B. The formulas of the reactants and the products are not correct.

    C. The equation is not balanced, so it does not show conservation of mass.

    D. The types of atoms present on the reactant side are not the same as those on the product side.

3. One way in which a nuclear reaction is different from a chemical reaction is that the identity of the atom itself might change as a result of the reaction. Which of the following will cause an atom to change into an atom of a different element?

    A. The atom gains or loses energy.

    B. The atom gains or loses protons.

    C. The atom gains or loses neutrons.

    D. The atom gains or loses electrons.

Name _____ Date _____

Unit 4

4. The image below represents energy changes that occur during a chemical reaction.

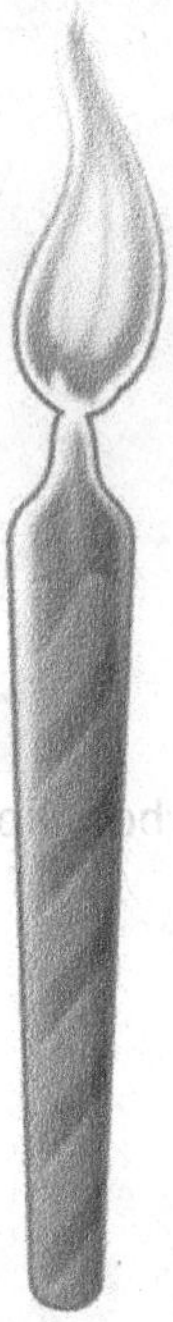

Based on the image, what can you infer?

A. The reaction is exothermic.

B. The reaction is endothermic.

C. The reaction destroys energy.

D. The reaction does not require energy.

5. Why does decreasing the temperature of a reaction decrease the rate at which the reaction occurs?

A. Lower temperatures decrease the amount of catalyst present.

B. Lower temperatures decrease the number of reactant particles.

C. Lower temperatures cause particles to move more slowly and collide less often.

D. Lower temperatures decrease the sizes of the reactant particles so they collide less often.

6. The following appears in a chemical equation:

$$3C_2H_4$$

How many atoms of carbon are represented by this formula?

A. 2

B. 3

C. 5

D. 6

Name _____ Date _____

Unit 4

7. The following chemical equations represent three different chemical reactions.

   $$3NaOH + H_3PO_4 \rightarrow Na_3PO_4 + 3H_2O$$
   $$2Mg + O_2 \rightarrow 2MgO$$
   $$Pb(NO_3)_2 + 2NaI \rightarrow PbI_2 + 2NaNO_3$$

   Based on these equations, which statement is true?

   A. Mass is lost when 2Mg and O_2 become 2MgO.

   B. Because Mg and O_2 are both elements, no bonds were broken during the formation of MgO.

   C. The chemical properties of $Pb(NO_3)_2$ and NaI differ from the chemical properties of PbI_2 and $NaNO_3$.

   D. The chemical properties of NaOH and Na_3PO_4 are similar, and the chemical properties of H_3PO_4 and H_2O are similar.

8. What property of carbon is the primary reason it can form so many different types of compounds?

   A. A carbon atom contains six protons in its nucleus.

   B. Each carbon atom contains four valence electrons.

   C. Certain carbon isotopes are capable of radioactive decay.

   D. Carbon exists in nature in solid, liquid, and gaseous states.

9. Helium-3 undergoes gamma decay. How does the release of gamma radiation affect an atom?

   A. It increases the mass number by 1.

   B. It decreases the mass number by 2.

   C. It increases the atomic number by 1.

   D. It has no effect on mass number or atomic number.

10. What is the cause of the large amount of energy produced in a nuclear fission reaction?

    A. The nucleus of a large atom splits into two smaller nuclei.

    B. The nucleus of a large atom absorbs electrons from smaller atoms.

    C. Nuclei of smaller atoms form bonds that bring them closer together.

    D. Nuclei of smaller atoms combine to form a new, more massive nucleus.

11. The diagram shows the structural formula for hexane.

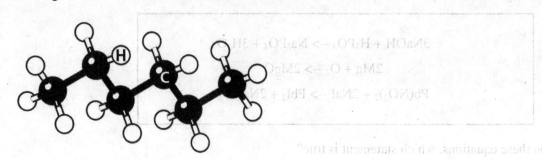

Based on the diagram, which of the following is the chemical formula for hexane?

A. C_3H_6

B. C_4H_8

C. C_6H_6

D. C_6H_{14}

12. Which of the following statements accurately describes the kinds of elements that can undergo radioactive decay?

A. All elements undergo radioactive decay.

B. Elements with unstable nuclei undergo decay.

C. Only elements created in laboratories undergo decay.

D. Elements with more neutrons than protons undergo decay.

Name _____ Date _____

Unit 4

Critical Thinking
Answer the following questions in the space provided.

13. One class of organic compounds is organic acids.

    What is an organic acid?

    _____
    _____

    What is a carboxyl group? Explain how you could use a diagram to represent a carboxyl group.

    _____
    _____
    _____

    What are two examples of organic acids?

    _____
    _____

Name _____ Date _____ Date _____

Unit 4

Extended Response
Answer the following questions in the space provided.

14. The diagram shows a helium nucleus that formed by nuclear fusion.

What are the mass number and charge of this nucleus?

_____
_____

Explain what happens during nuclear fusion.

_____
_____

How is mass related to energy in this reaction?

_____
_____
_____

Give an example of a similar reaction that occurs in nature.

_____

Solutions, Acids, and Bases

Choose the letter of the best answer.

1. The following diagram places sodium hydroxide on the pH scale.

   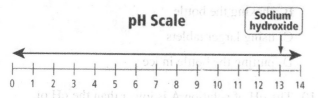

   Based on this diagram, which of these phrases describes sodium hydroxide?

   A. highly basic

   B. highly acidic

   C. slightly basic

   D. slightly acidic

2. Which of the following forms during a neutralization reaction?

   A. a salt

   B. a base

   C. an acid

   D. a metal

3. A scientist makes a solution that contains 80 g of sugar per 200 mL of solution. Which concentration describes the solution?

   A. 40%

   B. 80%

   C. 120%

   D. 280%

4. A colloid, a suspension, and a solution are placed in the containers below. Then a light is shined on the containers as shown.

   Which choice names the substance in each container, from left to right?

   A. solution, colloid, suspension

   B. suspension, solution, colloid

   C. colloid, suspension, solution

   D. solution, suspension, colloid

5. Which property of acids and bases do most acid-base indicators use?

   A. Acids and bases can react together to form salts.

   B. Acids and bases give off specific tastes that are easy to recognize.

   C. Acids and bases can react with certain objects to wear them away.

   D. Acids and bases react with certain compounds to cause a change in color.

Name _____ Date _____   Unit 5

6. Which of the following is a common effect of acid rain?

   A. The volume of oceans is increased.

   B. The temperature of rivers is increased.

   C. The pH of lakes and streams is reduced.

   D. The amount of water on Earth is reduced.

7. Which of these is a common use of salts?

   A. as antacids

   B. as detergents

   C. as roadway de-icers

   D. as components in fertilizer

8. The diagram below shows an ionic compound placed in water.

   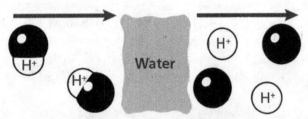

   What can you conclude about the ionic compound based on the diagram?

   A. It is a salt.

   B. It is a base.

   C. It is an acid.

   D. It is neutral.

9. Hikers often add iodine tablets to a bottle of mountain water to make it safe to drink. Which of the following will cause the tablet to dissolve more quickly in the water?

   A. adding more water

   B. shaking the bottle

   C. using larger tablets

   D. putting the bottle in ice

10. The pH of solution A is lower than the pH of solution B. What does this indicate about the solutions?

    A. The volume of solution A is less than the volume of solution B.

    B. Solution A has a greater hydrogen ion concentration than solution B.

    C. The particles in solution A are smaller than the particles in solution B.

    D. The average speed of the particles in solution A is less than in solution B.

Name _____ Date _____

Unit 5 Lesson 1
Lesson Quiz

Solutions

Choose the letter of the best answer.

1. Which reason explains why gas bubbles are released when a carbonated beverage is opened?

   A. Decreasing the pressure increases the rate of solution.

   B. The solubility of a gas in a liquid decreases with decreased pressure.

   C. Exposing a liquid to air causes it to become a saturated solution.

   D. The solvent changes from a liquid to a gas when the beverage is opened.

2. How is a dilute solution different from a concentrated solution?

   A. The dilute solution has a liquid solvent.

   B. The dilute solution is at a lower temperature.

   C. The dilute solution has a smaller total volume.

   D. The dilute solution has less solute per volume of solvent.

3. Saltwater is an example of a solution. Compared to the salt, which term best describes the water in the solution?

   A. solute

   B. solvent

   C. mixture

   D. suspension

4. Sugar in jar A is added to water in jar B, as shown in the diagram below.

   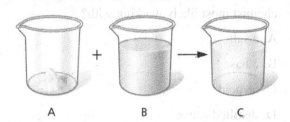

   Which statement is true about the substance in Jar C?

   A. The substance is a homogeneous mixture.

   B. The sugar remains separate from the water.

   C. The particles in the substance will scatter light.

   D. The substance can no longer be separated by physical means.

5. Which of the following will slow the rate at which sugar dissolves in water?

   A. cooling the water

   B. crushing the sugar

   C. increasing the amount of sugar

   D. stirring the sugar into the water

Name _____ Date _____

Unit 5 Lesson 2

Lesson Quiz

Acids, Bases, and Salts

Choose the letter of the best answer.

1. A chemist is working with a liquid substance that tastes bitter, feels slippery, and conducts an electric current. Which of the following is the chemist most likely working with?

   A. a salt

   B. a base

   C. an acid

   D. distilled water

2. How is a salt formed during a neutralization reaction?

   A. A hydrogen ion reacts with a hydroxide ion.

   B. An acid changes to a different state of matter.

   C. A metal ion takes the place of the hydrogen of an acid.

   D. The water molecule breaks into hydroxide and hydrogen ions.

3. What happens to an acid when it dissolves in water?

   A. It increases the number of hydroxide ions in water.

   B. It increases the number of hydrogen ions in water.

   C. It produces a salt that collects at the bottom of the container.

   D. It breaks water into individual atoms of hydrogen and oxygen.

4. The equation below describes a neutralization reaction.

   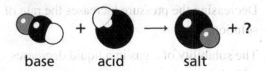

   base   acid   salt

   Which substance is missing from the product side of the equation?

   A. water

   B. oxygen

   C. hydrogen ion

   D. hydroxide ion

5. How is hydrochloric acid important to maintaining the health of the human body?

   A. It carries oxygen in the blood to where it is needed.

   B. It breaks down food so that the body can obtain nutrients.

   C. It removes wastes that are produced during normal processes.

   D. It kills pathogens so that the body can protect itself from disease.

Name _____ Date _____

Unit 5 Lesson 3
Lesson Quiz

Measuring pH

Choose the letter of the best answer.

1. How does absorbing carbon dioxide gas from the atmosphere most directly affect oceans?

   A. It lowers the pH of the water.

   B. It raises the temperature of the water.

   C. It reduces the volume of water in the oceans.

   D. It increases the salt concentration of the water.

2. Which of the following is indicated by the pH value of a solution?

   A. its hydrogen ion concentration

   B. its ammonium ion concentration

   C. its ability to undergo a chemical reaction

   D. its ratio of solute amount to solvent volume

3. How is the pH value of a solution related to its alkalinity?

   A. The lower the pH is, the higher the alkalinity is.

   B. The higher the pH is, the higher the alkalinity is.

   C. The pH value is unrelated to the solution's alkalinity.

   D. The pH value is equivalent to the solution's alkalinity.

4. Which of the following is a common cause of acid rain?

   A. oil spilled into tropical ocean waters

   B. gases released when plants make food

   C. emissions produced by burning fossil fuels

   D. warm water released from nuclear power plants

5. The diagram below identifies the locations of several substances on the pH scale.

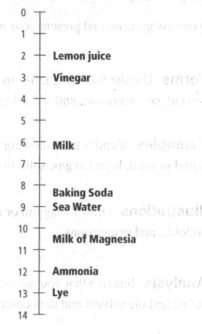

Human blood is slightly basic. Where would you add human blood to the diagram?

A. between ammonia and lye

B. between milk and baking soda

C. between lemon juice and vinegar

D. between milk of magnesia and ammonia

Name _____ Date _____

Unit 5 Lesson 1
Alternative Assessment

Solutions

Points of View: *What Is the Solution?*
Your class will work together to show what you've learned about solutions from several different viewpoints.

1. Work in groups as assigned by your teacher. Each group will be assigned to one or two viewpoints.

2. Complete your assignment, and present your perspective to the class.

 Terms Use the following terms to write a paragraph describing solutions: *solution, solute, solvent, concentration,* and *solubility*.

 Examples Identify the following types of solutions and provide examples: liquid in liquid, liquid in solid, liquid in gas, solid in solid, solid in gas, and gas in gas.

 Illustrations Draw a diagram or illustration that shows the difference between solutions, colloids, and suspensions.

 Analysis Imagine that you are making a solution of copper sulfate and water. Identify the solute and the solvent and then describe how you can increase its solubility.

 Models Imagine you are a molecule of a solute that will be used to make a solution. Reenact how you will be used to make a solution and demonstrate how temperature and pressure will affect the solubility of the solution.

Name _____ Date _____

Unit 5 Lesson 2
Alternative Assessment

Acids, Bases, and Salts

Climb the Ladder: *Salty with a Chance of Acids and Bases*
Complete the following to show what you have learned about acids, bases, and salts.

1. Work on your own, with a partner, or with a small group.
2. Choose one item from each rung of the ladder. Check your choices.
3. Have your teacher approve your plan.
4. Submit or present your results.

__ **Staying Neutral** Draw a diagram or illustration that shows a neutralization reaction. At the bottom, identify three salts and tell how each is commonly used.	__ **Pass the Salt!** Write a paragraph that describes a neutralization reaction. Give examples of salts and tell how each is commonly used.
__ **Base Hit** Write a song, poem, or paragraph that tells the properties of bases.	__ **Basic Drama** Write and perform a skit that demonstrates how bases break up in water.
__ **Acid Looks** Prepare a multimedia presentation that shows how acids break up in water.	__ **Acids By the Book** Write a booklet for third graders that identifies the properties of acids. Illustrate your booklet.

Name _____ Date _____

Unit 5 Lesson 3

Alternative Assessment

Measuring pH

Take Your Pick: *pH All Around Us*
Follow the steps below to show what you have learned about the pH scale and the role of acids and bases in our world.

1. Work on your own, with a partner, or with a small group.
2. Choose items below for a total of 10 points. Check your choices.
3. Have your teacher approve your plan.
4. Submit or present your results.

2 Points

_____ **The pH Scale** Draw a pH scale. Show where each of the following items belongs on the scale: pure water, normal rain water, lemon juice, baking soda, soap, and acid rain.

_____ **Difference in pH** If a solution changes from a pH of 9 to a pH of 5, what is the change in hydrogen ion concentration?

5 Points

_____ **Soil pH** Write a gardening column for a newspaper, explaining how the pH of soil affects the flowers of hydrangea plants and how soil pH can be adjusted to change a hydrangea's color. Before writing, research techniques used by gardeners to raise and lower soil pH.

_____ **Stomach pH** Create a pamphlet that could be handed out in medical offices to people who experience frequent problems with indigestion. The pamphlet should describe causes of indigestion, why someone might take antacid tablets, and how they affect the pH of the stomach.

_____ **Measuring pH** Make a poster that uses illustrations and captions to show three methods for measuring the pH of an unknown solution. The poster should identify which method provides the most accurate result.

8 Points

_____ **In Our Bodies** Identify three conditions that could cause a change in pH within a body system. Explain the cause of each change in pH, the effects on the person's health, and how the body restores pH back to a healthy level. Draw pictures or a diagram to illustrate your answers.

_____ **Acid Rain** Diagram the cause of acid rain and three ways it affects the world around you. Then, identify two ways that humans can reduce the amount of acid-causing chemicals released into the atmosphere.

Unit 5

Performance-Based Assessment Teacher Notes

Solubility and Temperature

Purpose In this activity, students observe the solubility of solids and gases in water at different temperatures, and draw conclusions about the relationship between solubility and temperature.

Time Period 45-60 minutes

Preparation Equip each activity station with the necessary materials.

Safety Tips Students should not eat or drink anything in the laboratory. Students should not heat glassware that is broken, chipped, or cracked. They should use tongs or heatproof gloves to handle heated glassware and other equipment. Clean up water spills immediately; spilled water is a slipping hazard. Never work with electricity near water; be sure the floor and all work surfaces are dry. Students should always wear heat-resistant gloves, goggles, and an apron when using a hot plate to protect their eyes and clothing. Never leave a hot plate unattended while it is turned on. Allow all equipment to cool before storing it. Students should tie back long hair, secure loose clothing, and remove loose jewelry.

Teaching Strategies This activity works best in groups of 3–4 students. For this activity, "boiling" is the point at which students first see bubbles form on the bottom of the beaker.

Scoring Rubric

Possible points	Performance indicators
0-30	Forming and testing hypothesis
0-40	Analyzing results
0-30	Drawing conclusions

Name _____ Date _____

Unit 5

Performance-Based Assessment

Solubility and Temperature

Objective
You have learned that both pressure and temperature affect the solubility of a solute in a solvent. In this activity, you will investigate how solids and gases dissolve in water at different temperatures.

Know the Score!
As you work through this activity, keep in mind that you will be earning a grade for the following:
- how well you form and test the hypothesis (30%)
- the quality of your analysis (40%)
- the clarity of your conclusions (30%)

Materials and Equipment
- beakers, 250 mL (2)
- graduated cylinder
- gloves, heat-resistant
- hot plate
- paper towels
- salt, 100 g
- spoon, (1/4 teaspoon)
- stirring rod or spoon
- water, cold, 200 mL (1 cup)

Safety Information
- Be sure to clean up water spills immediately because spilled water is a slipping hazard.
- Be careful if a glass beaker breaks.
- Tie back long hair and remove any loose jewelry.

Procedure

1. Form a hypothesis that explains how the temperature of a solvent will affect solubility.

   _____
   _____

2. Use your hypothesis to write a prediction about what will happen when you add salt to water of two temperatures.

   _____
   _____

3. Pour 100 mL of cold water into each empty beaker.

4. Wearing safety equipment, plug in and turn on the hot plate. Place one of the beakers on the hot plate.

5. When the water begins to boil, remove the beaker from the hot plate. Place it next to the other beaker.

Name _____ Date _____

Unit 5

6. To each beaker, add one spoonful of salt at a time while stirring. Continue to add salt until no more salt dissolves. You will know when no more salt will dissolve when you see salt fall to the bottom no matter how much you stir.

7. Record the amount of salt dissolved in the water. (Each spoonful is about 1.5 g.)

Amount of Dissolved Salt

Cold Water (g)	Hot Water (g)

Analysis

8. Did the cold water or the hot water dissolve more salt?
   _____
   _____

9. Compare the data you gathered to your prediction. Was your prediction correct? Explain your answer.
   _____
   _____
   _____

10. Compare the data you gathered to your hypothesis. Was your hypothesis supported? Explain your answer.
    _____
    _____
    _____

11. In this experiment, you worked with water, table salt, and salt water. Classify each of these materials as a solute, solvent, or solution. Then, explain why you classified them as such.
    _____
    _____
    _____
    _____

Name _____ Date _____

Unit 5: Solutions, Acids, and Bases

Unit 5
Unit Review

Vocabulary
Check the box to show whether each statement is true or false.

T	F	
☐	☐	1. An <u>acid</u> is any compound that increases the number of hydroxide ions when dissolved in water.
☐	☐	2. The <u>pH</u> value of a solution is a measure of the solution's hydronium ion concentration.
☐	☐	3. <u>Concentration</u> is the amount of a particular substance in a given quantity of a solution.
☐	☐	4. <u>Solubility</u> is the ability of one substance to dissolve in another at a given temperature or pressure.
☐	☐	5. A <u>salt</u> is an covalent compound that forms when a metal atom replaces the hydrogen of an acid.

Key Concepts
Read each question below, and circle the best answer.

6. Mylie tested the pH of ammonia and lemon juice. What was most likely her result?

   A. Both ammonia and lemon juice had a pH value less than 7.

   B. Both ammonia and lemon juice had a pH value more than 7.

   C. Ammonia had a pH value less than 7, and lemon juice had a value greater than 7.

   D. Ammonia had a pH value greater than 7, and lemon juice had a value less than 7.

7. In our bodies, stomach acids help break down food and release nutrients. Why do people take antacids sometimes?

   A. make their stomach more acidic

   B. make their stomach less basic

   C. neutralize some stomach acids

   D. release more nutrients from their food

Name _____ Date _____

Unit 5

8. Below is a scale used in characterizing a solution.

pH Scale

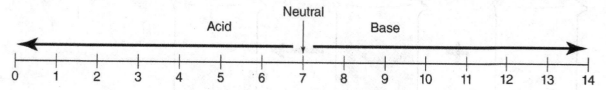

An acidic solution undergoes a 100 fold decrease in hydronium ion concentration. Which of the following intervals could represent this change?

A. 0–1
B. 2–4
C. 8–10
D. 9–12

9. Light from a flashlight is passed through liquids in two glasses. In Glass A, the light passes straight through. In Glass B, the light is scattered.

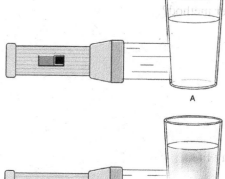

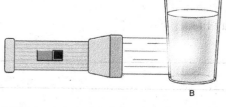

Which glass(es) might contain a solution?

A. Glass A
B. Glass B
C. both Glass A and Glass B
D. neither Glass A or Glass B

10. Why does stirring or crushing a solid solute make it dissolve faster in a liquid?

A. It increases the surface area of the solvent.

B. It increases the surface area of the solute.

C. It increases the temperature of the solvent.

D. It increases the pressure of the solvent.

11. What is a solution that has a relatively low amount of solute called?

A. saturated
B. concentrated
C. dilute
D. suspension

Name _____ Date _____

Unit 5

12. Alayna makes a solution as shown in the diagram below.

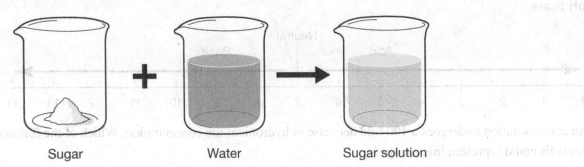

Sugar   Water   Sugar solution

What is the solute shown in this diagram?

A. sugar   C. beaker
B. water   D. sugar solution

13. Universal pH paper and acid-base indicators both can be used to determine the pH of a solution. What property of the indicator changes when pH is measured with either method?

A. color   C. number
B. shape   D. letter

Critical Thinking

Answer the following questions in the space provided.

14. Give an example of an acid and explain how it is used in everyday life.

_____
_____
_____

Give an example of a base and explain how it is used in everyday life.

_____
_____
_____

15. Salad dressing and apple juice are both examples of mixtures.

Salad dressing   Apple juice

Which one is a solution?

_____

Name _____ Date _____

Unit 5

What properties of the mixture did you consider in your answer?

_____
_____

How would filtering the solution affect it?

_____
_____

Connect ESSENTIAL QUESTIONS

Lessons 2 and 3

Answer the following question in the space provided.

16. How is acid rain formed?

_____
_____
_____

What happens to the pH of soil, lakes, ponds, and streams when acid rain falls?

_____

Give an example of how acid rain may affect a living organism.

_____
_____
_____

What could be done to reduce the amount of acid rain from falling on Earth?

_____
_____
_____

Name _____ Date _____

Unit 5

Unit Test A

Solutions, Acids, and Bases

Key Concepts
Choose the letter of the best answer.

1. The diagram below identifies several substances on the pH scale.

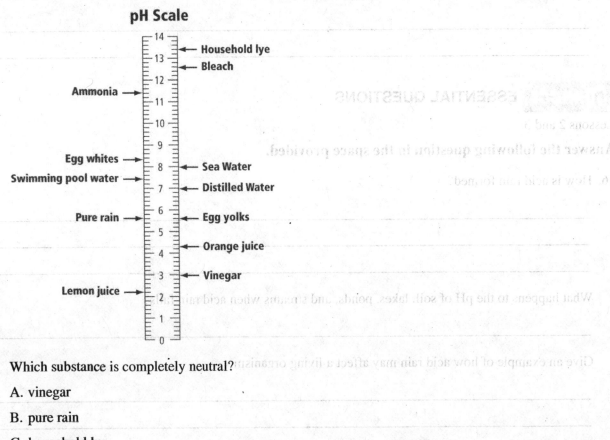

   Which substance is completely neutral?

   A. vinegar

   B. pure rain

   C. household lye

   D. distilled water

2. Since the 1700s, scientists have observed that the pH of ocean water is decreasing as the water absorbs carbon dioxide from the atmosphere. What does this indicate about carbon dioxide?

   A. It is salty.

   B. It is basic.

   C. It is acidic.

   D. It is neutral.

3. Vinegar is a solution of acetic acid in water. Which term best describes water in this solution?

   A. solute

   B. solvent

   C. solubility

   D. concentration

4. The graph shows how the solubility of different substances is affected by temperature.

   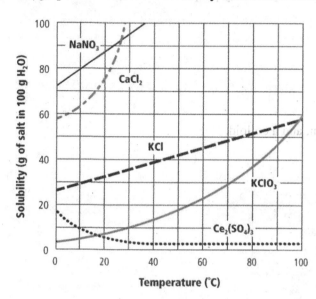

   For which substance does solubility show a decrease as temperature increases?

   A. KCl

   B. NaNO₃

   C. KClO₃

   D. Ce₂(SO₄)₃

5. Which of these substances might combine with water in the atmosphere to form acid rain?

   A. nitric acid

   B. ammonia

   C. citric acid

   D. hydrochloric acid

6. A man ate a tablet to neutralize too much acid in his stomach. What can you conclude about the tablet?

   A. It must have been a salt.

   B. It must have been sugar.

   C. It must have been a base.

   D. It must have been an acid.

7. Which of the following is an example of a salt commonly used by people?

   A. chalk

   B. bleach

   C. ammonia

   D. milk of magnesia

8. The diagram below shows a suspension, a solution, and a colloid.

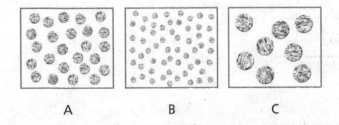

   Which property can you use to figure out which mixture is a colloid?

   A. the number of particles

   B. the size of the particles

   C. the color of the particles

   D. the temperature of the particles

Name _____ Date _____

Unit 5

9. A scientist is using bromthymol blue to analyze a solution. What is the scientist most likely doing?

   A. diluting the solution

   B. measuring the pH value

   C. determining the concentration

   D. separating the solute from the solvent

10. What happens to a base when it dissolves in water?

    A. It increases the number of hydroxide ions in the water.

    B. It increases the number of hydrogen ions in the water.

    C. It produces a salt that sinks to the bottom of the water.

    D. It breaks the water into individual atoms of hydrogen and oxygen.

11. How can a dilute solution of salt in water be made more concentrated?

    A. by heating it

    B. by shaking it

    C. by adding salt to it

    D. by adding water to it

12. Acids and bases are corrosive. Consequently, which of the following is a true statement?

    A. Acids and bases are essential nutrients in living things.

    B. Acids and bases create a protective coating on metals.

    C. Acids and bases are not found naturally in the environment.

    D. Acids and bases can react with and destroy other substances.

Name _____ Date _____

Unit 5

Critical Thinking
Answer the following questions in the space provided.

13. Substances can be described by their solubility.

    What is solubility?

    _____

    _____

    Using what you know about solubility, explain how a saturated solution can be dilute.

    _____

    _____

Name _____ Date _____

Unit 5

Extended Response
Answer the following questions in the space provided.

14. Two solutions are shown in the diagram below.

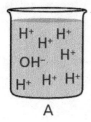

Which solution is acidic?
_____

How do you know?
_____
_____

Which solution is basic?
_____

How do you know?
_____
_____

Name _____ Date _____

Solutions, Acids, and Bases

Unit 5

Unit Test B

Key Concepts
Choose the letter of the best answer.

1. The diagram below identifies several substances on the pH scale.

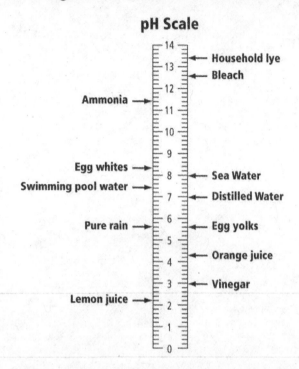

   Which statement about the identified substances is true?

   A. Bleach has a higher hydrogen ion concentration than sea water.

   B. Orange juice has a higher hydroxide ion concentration than pure rain.

   C. Egg whites have a higher hydroxide ion concentration than household lye.

   D. Egg yolks have a higher hydrogen ion concentration than egg whites.

2. The growth of hard corals in some parts of the ocean requires a pH range between 8.3 and 8.5. What might cause the pH to drop closer to 8.0?

   A. The water absorbs salt, such as sodium chloride.

   B. The water absorbs a base, such as calcium hydroxide.

   C. The water absorbs dirt particles from the ocean floor.

   D. The water absorbs an acid, such as carbon dioxide gas.

Name _____ Date _____

Unit 5

3. A student makes a solution of salt and water. What will happen if the student pours the solution through a filter?

   A. The salt will filter out of the solution.

   B. Only the water will be blocked by the filter.

   C. Both the salt and water will pass through the filter.

   D. The salt and water will undergo a chemical reaction.

4. The graph shows how the solubility of several substances is affected by temperature.

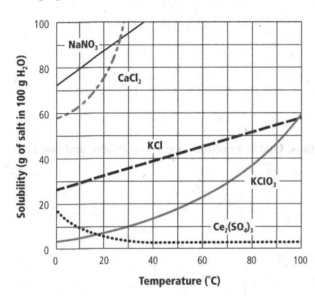

   By about how much does the solubility of KCl increase as the temperature is raised from 10°C to 70°C?

   A. about 10 grams per 100 g H_2O

   B. about 20 grams per 100 g H_2O

   C. about 30 grams per 100 g H_2O

   D. about 50 grams per 100 g H_2O

5. Which of the following human activities is most likely to result in acid rain?

   A. driving gasoline-burning vehicles

   B. spraying chemical pesticides on crops

   C. adding filters to smokestacks to trap carbon dioxide

   D. releasing ozone-destroying compounds into the atmosphere

Name _____ Date _____

Unit 5

6. Phenolphthalein is pink in the presence of a base and colorless when a base is not present. A student injected a solution of a base containing phenolphthalein into a lemon. When she cut the lemon open, there was no pink color inside. What must have happened?

   A. The acid in the lemon neutralized the base.

   B. The base changed into an acid in the lemon.

   C. The phenolphthalein stopped working inside the lemon.

   D. The lemon absorbed the base but not the phenolphthalein.

7. How do ants use formic acid to stay alive?

   A. They use it to identify food.

   B. They use it to feed their young.

   C. They use it to break down food.

   D. They use it to defend themselves.

8. The diagram below shows several different mixtures. One is a colloid, one is a solution, and one is a suspension.

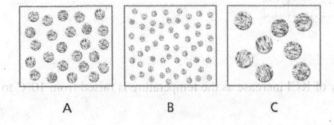

   Based on what you know about mixtures, which of the following statements is true?

   A. The container on the left contains a colloid.

   B. The container on the left contains a solution.

   C. The container on the right contains a solution.

   D. The container in the center contains a suspension.

Name _____ Date _____   Unit 5

9. The pH of a sample of acid rain is 4 and the pH of orange juice is 3. The hydrogen ion concentration of acid rain is 1,000 times greater than the hydrogen ion concentration of pure water. How many times greater is the hydrogen ion concentration of orange juice?

   A. 100

   B. 1,000

   C. 10,000

   D. 100,000

10. Sodium hydroxide is a strong base. When sodium hydroxide dissolves in water, which of the following is produced?

    A. H^+ ions

    B. OH^- ions

    C. salt crystals

    D. water molecules

11. What is the concentration of a solution made up of 25 g of sodium chloride in 125 g of solution?

    A. 0.2%

    B. 5%

    C. 20%

    D. 150%

12. Which product is formed when acids react with most metals?

    A. water

    B. oxygen gas

    C. hydrogen gas

    D. hydroxide ions

Name _____ Date _____

Unit 5

Critical Thinking
Answer the following questions in the space provided.

13. A scientist dissolves sodium chloride in water at 25°C and 1 atm.

    The scientist finds that the solubility of sodium chloride in water under these conditions is 357 g/L. In your own words, explain what this means.

    _____

    _____

    The scientist continues to add sodium chloride to the solution. What will happen? Explain why.

    _____

    _____

Name _____ Date _____

Unit 5

Extended Response
Answer the following questions in the space provided.

14. Two solutions are shown in the diagram below.

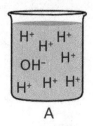
A

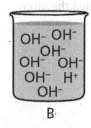
B

Give an example of a substance that might have been dissolved in water to form Solution A?
_____

Give an example of a substance that might have been dissolved in water to form Solution B?
_____

Explain why you chose each substance.
_____
_____

What will most likely happen if the two solutions are mixed?
_____
_____

Name _____ Date _____

Module H

End-of-Module Test

Matter and Energy

Choose the letter of the best answer.

1. Which observation is a sign of a chemical change?

    A. A rotting potato gives off a bad smell.

    B. A melting block of ice leaves a large puddle.

    C. A cloud changes shape when blown by wind.

    D. A plaster statue breaks when it falls onto the floor.

2. A British scientist named J. J. Thomson discovered the electron in 1897. He suggested that the atom looked like the model below, with electrons in a mass, or cloud, of positive charge.

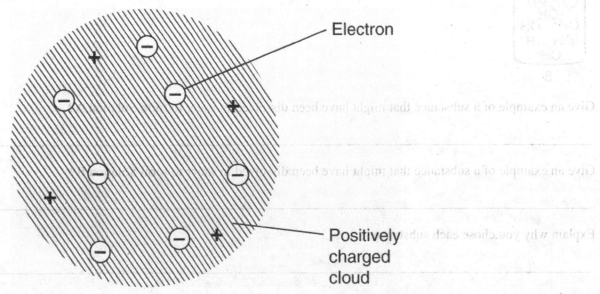

    How was the Thomson model different from the Bohr model that followed it?

    A. The Bohr model did not include positive charges in the atom.

    B. The Bohr model showed electrons surrounding the positive nucleus.

    C. The Bohr model consisted of a single, solid mass with no smaller particles.

    D. The Bohr model showed neutrons in a cloud surrounding the negative nucleus.

3. Which statement is supported by the atomic theory?

    A. Atoms combine to make all the substances on Earth.

    B. Atoms are made up of smaller particles called molecules.

    C. Atoms are easily destroyed when matter is cooled or heated.

    D. Atoms come in different sizes, but most are visible using a simple hand lens.

Name _____ Date _____

Module H

4. Which energy transfer takes place when water freezes to form ice crystals in the atmosphere?

   A. No energy is transferred.

   B. Energy is transferred from the air into the water.

   C. Energy is transferred from the water into the air.

   D. Energy is transferred from the water into the ground.

5. Leah used a triple-beam balance to measure the mass of a beaker of water.

   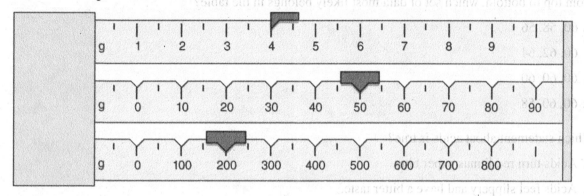

   Which value for mass should she record in her science notebook?

   A. 4 grams

   B. 250 grams

   C. 254 grams

   D. 254.5 grams

6. Zinc (Zn) is located directly to the left of gallium (Ga) on the periodic table. The atomic number of gallium is 31. What is the atomic number of zinc?

   A. 13

   B. 30

   C. 31

   D. 32

7. Fluorine has 7 valence electrons. How many more electrons does it need to achieve a full outermost energy level?

   A. 1

   B. 3

   C. 7

   D. 10

Name _____ Date _____

Module H

8. A student melts 60 g of ice in a sealed metal container that can withstand high pressure. He then heats the liquid water until all of it boils. He made this table to use in recording his results.

State	Mass (g)
Solid	
Liquid	
Gas	

From top to bottom, which set of data most likely belongs in the table?

A. 60, 58, 56

B. 60, 62, 64

C. 60, 60, 60

D. 60, 60, 58

9. Which statement about acids is true?

A. Acids turn red litmus paper blue.

B. Acids feel slippery and have a bitter taste.

C. Acids donate hydrogen ions to water when they dissolve.

D. Acids are ionic compounds formed when a metal atom replaces hydrogen.

10. A scientist is investigating long-chain carbon molecules composed of repeating structural units. Which of the following might the scientist be investigating?

A. water

B. starches

C. acid rain

D. radioactive tracers

11. A student is going to compress the object shown in the diagram.

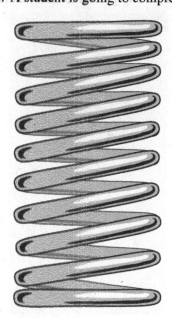

Which form of energy will the object gain when it is compressed?

A. chemical

B. electrical

C. mechanical

D. electromagnetic

12. Which statement is true of the carbon atoms that make up a diamond?

A. They are always vibrating in place.

B. They can slip past one another slowly.

C. They vibrate only when they are heated.

D. They are locked in place and do not move at all.

13. According to the kinetic theory of matter, how many particles that make up a substance are constantly in motion?

A. none of the particles

B. fewer than half of the particles but more than none

C. more than half of the particles but not all

D. all of the particles

Name _____ Date _____

Module H

14. When a hand-held fan is turned on, the blades spin. Which diagram shows the energy conversion that is required to make the fan work?

A. Electrical → Chemical → Light

B. Electrical → Chemical → Kinetic

C. Chemical → Electrical → Light

D. Chemical → Electrical → Kinetic

15. In which state(s) of matter are particles constantly moving?

A. gases only

B. liquids only

C. gases and liquids only

D. gases, liquids, and solids

Name _____ Date _____

Module H

16. The entry for one element in the periodic table is shown below.

   | 20 |
   | Ca |
   | Calcium |
   | 40.078 |

   What is the element's chemical symbol?

   A. 20

   B. Ca

   C. 40.078

   D. Calcium

17. Calcium oxide (CaO) is slightly soluble in water, whereas sodium chloride (NaCl) is highly soluble. What does this say about the solubility of the two compounds?

   A. Sodium chloride can dissolve more calcium oxide than water can.

   B. More sodium chloride than calcium oxide will dissolve in water.

   C. Sodium chloride makes the water acidic, whereas calcium oxide makes the water basic.

   D. Sodium chloride will sink to the bottom of a sample of water, whereas calcium chloride will float to the top.

18. Which type of energy involves harnessing heat produced within Earth to generate electricity?

   A. solar energy

   B. nuclear energy

   C. geothermal energy

   D. hydroelectric energy

Name _____ Date _____

Module H

19. Students put the same amount of water into four beakers and warm them to 60°C. They then place each beaker in different locations in the classroom. After 5 min, the students measure the temperature of each beaker of water. The results are shown in the bar graph below.

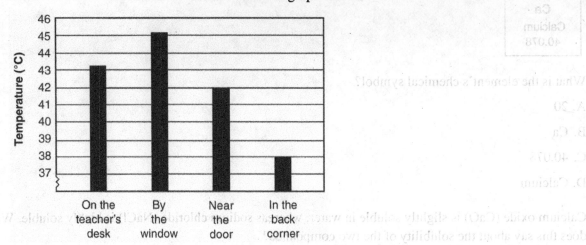

At which location do the particles of water move the slowest after 5 min?

A. near the door

B. by the window

C. in the back corner

D. on the teacher's desk

20. Which process can occur when the particles in a sample of matter experience an increase in kinetic energy?

A. freezing

B. deposition

C. evaporation

D. condensation

21. The density of aluminum is 2.7 g/cm³. What is the volume of a piece of aluminum if its mass is 8.1 grams?

A. 3.0 cm³

B. 2.7 cm³

C. 0.33 cm³

D. 21.9 cm³

Name _____ Date _____

Module H

22. In the laboratory, Austin is given a mixture of iron filings, sand, and salt. To separate the mixture, Austin uses a magnet, boiling water, and a filter. Which statement is true about the process Austin uses to separate this mixture?

   A. Austin uses only physical changes to separate the mixture's components.
   B. Austin uses only chemical changes to separate the mixture's components.
   C. Austin uses both physical and chemical changes to separate the mixture's components.
   D. Austin uses neither physical nor chemical changes to separate the mixture's components.

23. The diagram shows the formation of water.

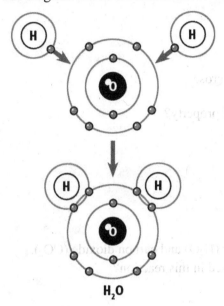

   Which of the following occurs when water forms?

   A. Two hydrogen atoms are destroyed to form atoms of water.
   B. One oxygen atom shares electrons with two hydrogen atoms.
   C. Two hydrogen atoms each give an electron to an oxygen atom.
   D. One oxygen atom gives an electron to each of two hydrogen atoms.

24. Alkalosis is a condition in which the blood pH rises too high. What causes the pH to rise?

   A. The blood becomes more basic.
   B. Water is absorbed into the blood.
   C. Salts are not removed from the blood.
   D. The level of acids increases in the blood.

End-of-Module Test
© Houghton Mifflin Harcourt Publishing Company

Name _____ Date _____

Module H

25. Which of these statements best describes physical properties?

   A. Physical properties behave identically for all matter under the same conditions.

   B. Physical properties can be observed without changing the identity of a substance.

   C. Physical properties are observed by seeing how a substance reacts with other substances.

   D. Physical properties cause atoms and molecules to change structure when substances are mixed.

26. Sucrose is another name for table sugar. Sucrose is a compound made from the elements carbon, hydrogen, and oxygen. Which statement best describes the properties of sucrose?

   A. They are exactly like the properties of carbon.

   B. They are exactly like the properties of oxygen.

   C. They are exactly like the properties of hydrogen.

   D. They are different from the properties of the elements in sucrose.

27. Which of these statements describes an example of a chemical property?

   A. A silver statue begins to tarnish.

   B. A painter coats a building with red paint.

   C. A freshly waxed floor has a bright shine.

   D. A metal turns to liquid at a certain temperature.

28. The fuel butane (C_4H_{10}) reacts with oxygen (O_2) to form water (H_2O) and carbon dioxide (CO_2). Which of the following best describes the carbon atoms involved in this reaction?

   A. The number of carbon atoms does not change.

   B. Three carbon atoms are formed in the reaction.

   C. Three carbon atoms are destroyed in the reaction.

   D. The arrangement of carbon atoms does not change.

29. An atom undergoes radioactive decay and emits an alpha particle. What happens to the mass number of the atom?

   A. It stays the same.

   B. It increases by two.

   C. It decreases by one.

   D. It decreases by four.

Name _____ Date _____

Module H

30. The graph below represents energy changes that occur during a chemical reaction.

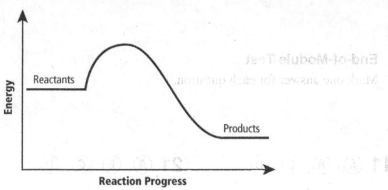

What can you conclude based on the graph?

A. The reaction absorbs energy.

B. The reaction releases energy.

C. The reaction destroys energy.

D. The reaction does not involve energy.

Name _____   Date _____

Answer Sheet

PLEASE NOTE
• Use only a no. 2 pencil
• Example: Ⓐ ● Ⓒ Ⓓ
• Erase changes COMPLETELY.

End-of-Module Test
Mark one answer for each question.

1 Ⓐ Ⓑ Ⓒ Ⓓ   11 Ⓐ Ⓑ Ⓒ Ⓓ   21 Ⓐ Ⓑ Ⓒ Ⓓ

2 Ⓐ Ⓑ Ⓒ Ⓓ   12 Ⓐ Ⓑ Ⓒ Ⓓ   22 Ⓐ Ⓑ Ⓒ Ⓓ

3 Ⓐ Ⓑ Ⓒ Ⓓ   13 Ⓐ Ⓑ Ⓒ Ⓓ   23 Ⓐ Ⓑ Ⓒ Ⓓ

4 Ⓐ Ⓑ Ⓒ Ⓓ   14 Ⓐ Ⓑ Ⓒ Ⓓ   24 Ⓐ Ⓑ Ⓒ Ⓓ

5 Ⓐ Ⓑ Ⓒ Ⓓ   15 Ⓐ Ⓑ Ⓒ Ⓓ   25 Ⓐ Ⓑ Ⓒ Ⓓ

6 Ⓐ Ⓑ Ⓒ Ⓓ   16 Ⓐ Ⓑ Ⓒ Ⓓ   26 Ⓐ Ⓑ Ⓒ Ⓓ

7 Ⓐ Ⓑ Ⓒ Ⓓ   17 Ⓐ Ⓑ Ⓒ Ⓓ   27 Ⓐ Ⓑ Ⓒ Ⓓ

8 Ⓐ Ⓑ Ⓒ Ⓓ   18 Ⓐ Ⓑ Ⓒ Ⓓ   28 Ⓐ Ⓑ Ⓒ Ⓓ

9 Ⓐ Ⓑ Ⓒ Ⓓ   19 Ⓐ Ⓑ Ⓒ Ⓓ   29 Ⓐ Ⓑ Ⓒ Ⓓ

10 Ⓐ Ⓑ Ⓒ Ⓓ   20 Ⓐ Ⓑ Ⓒ Ⓓ   30 Ⓐ Ⓑ Ⓒ Ⓓ

Test Doctor

Unit 1 Matter
Unit Pretest
1. D   5. C   9. A
2. D   6. A   10. A
3. D   7. B
4. C   8. A

1. **D**
   A is incorrect because 21 is the sum of the three values. The three values must be multiplied to get the volume.
   B is incorrect because 30 cm² is the area of one end, and not the volume.
   C is incorrect because 60 cm² is the area of one face, and not the volume.
   D is correct because the volume of a rectangle is equal to its length times width times height. In this example: 5 cm × 10 cm × 6 cm = 300 cm³.

2. **D**
   A is incorrect because evaporation is a physical property, not a chemical property.
   B is incorrect because mass is a physical property, not a chemical property.
   C is incorrect because state of matter is a physical property, not a chemical property.
   D is correct because reactivity is a chemical property.

3. **D**
   A is incorrect because a compound contains more than one type of element.
   B is incorrect because, although hydrogen exists as two atoms bonded together, the two atoms are of the same type.
   C is incorrect because water was the product.
   D is correct because a substances combines with oxygen when it burns. Water is therefore a compound made up of more than hydrogen.

4. **C**
   A is incorrect because freezing is a change of state, which is a physical change rather than a chemical change.
   B is incorrect because rolling a metal bar into a flat sheet is a change in shape and size, which is a physical change rather than a chemical change.
   C is correct because when vinegar and baking soda are mixed, a new substance is formed. This process is a chemical change.
   D is incorrect because forming a mixture of sand, water, and salt does not change any component's chemical composition. This process is a physical change, not a chemical change.

5. **C**
   A is incorrect because between points 1 and 2, the temperature of the ice is 0°C and the ice is melting.
   B is incorrect because between points 2 and 3, the temperature of the water is changing, so a change of state cannot be taking place.
   C is correct because between points 3 and 4, the temperature of the water is 100°C and the water is boiling.
   D is incorrect because between points 4 to 5, the temperature of the water vapor is changing, so a change of state cannot be taking place.

6. **A**
   A is correct because although mass is conserved in a chemical reaction, the evolution of a gas would explain the decrease in mass.
   B is incorrect because the evolution of a gas, not vinegar's evaporation, caused the decrease in mass.
   C is incorrect because mixtures are not less massive than their parts.
   D is incorrect because when baking soda and vinegar are mixed, mass is not destroyed. The process observes the Law of Conservation of Mass.

7. **B**
   A is incorrect because air is a gas, so the particles are far apart and move independently.
   B is correct because the air particles are free to move far apart and take up space.
   C is incorrect because it is the amount of space between the

particles that gets bigger, not the particles themselves.

D is incorrect because there is little attraction between the particles and the balloon, and such an attraction would not increase the volume.

8. **A**

A is correct because density is a ratio of mass and volume that stays the same regardless of the amount of a substance.

B is incorrect because mass is a measure of the amount of matter, so it does change when the amount of the substance changes.

C is incorrect because volume is a measure of how much room the substance takes up, so it does change when the amount of the substance changes.

D is incorrect because weight is a measure of the force of gravity acting on the mass. Because mass changes as the amount of the substance changes, so does the weight.

9. **A**

A is correct because 0°C is both the freezing point and melting point of water.

B is incorrect because ice melts at this temperature on the Fahrenheit scale; however, these temperatures are measured on the Celsius scale.

C is incorrect because 100°C is the boiling point of water.

D is incorrect because water boils at this temperature on the Fahrenheit scale.

10. **A**

A is correct because the parts of this mixture will separate over time.

B is incorrect because a gelatin dessert is a colloid.

C is incorrect because whipped cream is a colloid.

D is incorrect because apple juice is a solution.

Lesson 1 Quiz
1. A   4. A
2. B   5. D
3. D

1. **A**

A is correct because matter is anything that has mass and takes up space.

B is incorrect because some matter, such as air, has mass and takes up space, but is not visible.

C is incorrect because matter can also exist as a liquid or gas.

D is incorrect because liquids and gases change their shape depending on the container they are placed in.

2. **B**

A is incorrect because mass is a measure of the amount of matter in an object, which does not change if the object is moved to the moon.

B is correct because the coin weighs less on the moon, where the gravitational force is weaker.

C is incorrect because the volume of the coin is the amount of space it takes up. The volume is the same whether the coin is on the moon or on Earth.

D is incorrect because the density is a physical property determined by the type of matter in the coin. The makeup of the coin does not change on the moon.

3. **D**

A is incorrect because 400 cm^3 is the area of the base, but not the volume.

B is incorrect because the volume is obtained by multiplying length times width times height.

C is incorrect because all three measurements must be multiplied to obtain the volume.

D is correct because 40 cm × 10 cm × 5 cm = 2,000 cm^3.

4. **A**

A is correct because mass is a measure of the amount of matter in an object. The type of matter in the ice cube is water.

B is incorrect because weight is dependent upon gravity, whereas mass is not.

C is incorrect because the space the ice cube occupies is its volume.

D is incorrect because this product represents the volume of a rectangular solid.

5. **D**

A is incorrect because milk has a greater density than gasoline does, so milk occupies a smaller volume for a given mass.

Answer Key

B is incorrect because water has a greater density than gasoline does, so water occupies a smaller volume for a given mass.

C is incorrect because mercury has the greatest density, so it occupies the least volume for a given mass.

D is correct because gasoline has the least density, so it occupies the greatest volume for a given mass.

Lesson 2 Quiz
1. D   4. A
2. A   5. D
3. C

1. D

A is incorrect because chemical composition is a chemical property, not a physical property.

B is incorrect because flammability is a chemical property, not a physical property.

C is incorrect because reactivity with oxygen is a chemical property, not a physical property.

D is correct because density is a physical property.

2. A

A is correct because the ability to burn is a chemical property of paper.

B is incorrect because the paper's ability to be crumpled is a physical property, not a chemical property.

C is incorrect because the paper's inability to attract a magnet is a physical

property, not a chemical property.

D is incorrect because the paper's inability to conduct electricity is a physical property, not a chemical property.

3. C

A is incorrect because mass is the amount of matter in a sample. Mass depends on the size of the sample.

B is incorrect because volume describes the amount of space a sample takes up. Volume changes with the size of the sample.

C is correct because density is a characteristic physical property.

D is incorrect because flammability is a chemical property, not a physical property.

4. A

A is correct because, although the objects are made from the same metal, the masses of the four objects are not the same.

B is incorrect because the objects are made from the same metal, so their magnetism is the same.

C is incorrect because the objects are made from the same metal, so, their specific heat is the same.

D is incorrect because the objects are made from the same metal, so their electrical conductivity is the same.

5. D

A is incorrect because color is a physical property. It can be

observed without changing the nature of the object.

B is incorrect because texture is a physical property, not a chemical property.

C is incorrect because density is a physical property. It does not depend on the object's ability to change into another substance.

D is correct because reactivity with acid is a chemical property.

Lesson 3 Quiz
1. A   4. B
2. A   5. B
3. D

1. A

A is correct because changing the shape of an object by cutting it does not change its chemical makeup.

B is incorrect because the digestion of food is a chemical change. Food is broken down into different substances that enable them to be used by the body.

C is incorrect because the reaction between sodium and chlorine is a chemical change that produces different substances with different properties.

D is incorrect because the reaction of sodium and water is a chemical change that produces different substances with different properties.

2. A

A is correct because rusting results in the formation of new substances with different chemical properties.

B is incorrect because cooling is a physical property, not a chemical change.

C is incorrect because metal's expansion during heating is a physical change, not a chemical change.

D is incorrect because the breaking of glass in a window is a physical change, not a chemical change.

3. D

A is incorrect because a change in volume does not explain why the mass of the ash is less than the mass of the paper. A change in volume should not affect the amount of matter present.

B is incorrect because according to the law of conservation of mass, mass cannot be created or destroyed during a chemical change.

C is incorrect because the mass of the ash can be accurately determined with a balance.

D is correct because some of the mass appears to have been lost due to the gases that escaped. Adding the mass of the escaped gas to the mass of the other products should result in a value equal to that of the mass of the starting materials.

4. B

A is incorrect because boiling liquid involves a change in state as the liquid turns to steam. A change of state is a physical change rather than a chemical change.

B is correct because metal rusting involves a change in chemical composition, which is a chemical change.

C is incorrect because alcohol evaporating involves a change in state, which is a physical change rather than a chemical change.

D is incorrect because a piece of wood shrinks as it dries, which is simply a change in size. Therefore, this process is a physical change, not a chemical change.

5. B

A is incorrect because tearing paper changes its shape, not its mass.

B is correct because tearing paper changes its shape. This is a physical change.

C is incorrect because tearing paper changes its shape. This is a physical change, not a chemical change.

D is incorrect because tearing paper changes its shape. This is a physical change, not a change in energy.

Lesson 4 Quiz

1. B   4. D
2. D   5. D
3. B

1. B

A is incorrect because a mixtures always contain more than one type of atom.

B is correct because an element is made up of one type of atom.

C is incorrect because a molecule can be made up of more than one type of atom.

A water molecule is an example of a molecule made up of two types of atom (hydrogen and oxygen).

D is incorrect because a compound is made up of two or more types of atoms that are chemically combined.

2. D

A is incorrect because table salt is a compound.

B is incorrect because pure water is a compound.

C is incorrect because whole milk is a colloid, which is a heterogeneous mixture.

D is correct because maple syrup is the same throughout.

3. B

A is incorrect because milk is a colloidal mixture.

B is correct because brass is a homogeneous mixture of two metals.

C is incorrect because mercury is a liquid element.

D is incorrect because concrete is a heterogeneous mixture.

4. D

A is incorrect because each sphere in the diagram represents an atom. Product C consists of combinations of atoms.

B is incorrect because a mixture contains different types of elements and/or compounds that are physically combined. Product C consists of atoms that are chemically combined and cannot be separated by physical means.

Answer Key

C is incorrect because the dark-colored spheres represent one kind of element and the light-colored spheres represent another kind of element. Product C is made up of a combination of two different elements.

D is correct because a compound contains different types of atoms that are chemically combined.

5. D

A is incorrect because carbon is an element.

B is incorrect because chlorine is an element.

C is incorrect because uranium is an element.

D is correct because ammonia is a compound containing nitrogen and hydrogen.

Lesson 5 Quiz
1. C   4. A
2. A   5. B
3. D

1. C

A is incorrect because the atoms of an element remain the same despite changes in state.

B is incorrect because mass is conserved during a change of state.

C is correct because solids, liquids, and gases differ in the motion of their particles.

D is incorrect because a change of state does not change the identity of the particles.

2. A

A is correct because there is little attraction between the particles in a gas, so they move freely in all directions.

B is incorrect because ice is not a state of matter.

C is incorrect because liquids do not take the volume of their containers.

D is incorrect because solids do not take the shape or the volume of their containers.

3. D

A is incorrect because particles generally slide past one another in a liquid. Frost involves water in the solid state.

B is incorrect because deposition is a change from a gas to a solid. Particles sliding past each other do not describe a gas.

C is incorrect because particles do not gain more freedom when a gas changes into a solid.

D is correct because the particles of water vapor move quickly in all directions whereas the particles of ice only vibrate in place.

4. A

A is correct because freezing makes water into a solid, and water particles become locked in place.

B is incorrect because water particles in an ice cube still vibrate.

C is incorrect because water particles do not change size when they freeze.

D is incorrect because water particles cannot slip past one another in the solid state.

5. B

A is incorrect because only the particles of a solid are locked into position.

B is correct because particles of matter are in motion in each state. However, they experience greater motion in gases than in solids.

C is incorrect because solids do not take the shape of their containers.

D is incorrect because the volume of solids and liquids is constant, but gases fill their containers.

Lesson 6 Quiz
1. C   4. D
2. C   5. A
3. D

1. C

A is incorrect because evaporation occurs when a liquid changes to a gas. The mass of the solid and gas are the same.

B is incorrect because the change of a solid directly to a gas is sublimation, not freezing. A solid must absorb energy to sublimate.

C is correct because when a solid sublimes, it changes directly from a solid to a gas without first becoming a liquid. The kinetic energy of the gas particles is greater than the kinetic energy of the solid particles.

D is incorrect because the change of a solid directly to a gas is sublimation, not deposition. Mass is not lost during a change of state.

2. **C**

A is incorrect because mass is not lost during the melting process.

B is incorrect because a solid does not lose mass during a change of state.

C is correct because in a closed system, no matter is lost during a change of state.

D is incorrect because a liquid formed from a melted solid does not gain mass during the change.

3. **D**

A is incorrect because droplets on the grass in the morning is dew, which forms through a different process than deposition.

B is incorrect because evaporation is the change from a liquid to a gas, and dew is a liquid.

C is incorrect because sublimation is the change from a solid to a gas. A liquid must form for dew to appear on the grass.

D is correct because when energy is removed from gas particles, the attraction between the particles is great enough for them to be held together as a liquid, as seen in condensation.

4. **D**

A is incorrect because sublimation and melting result from an increase in the kinetic energy of the particles of a substance.

B is incorrect because melting and evaporation result from an increase in the kinetic energy of the particles of a substance.

C is incorrect because substances must absorb energy for sublimation, melting, and boiling to occur.

D is correct because substances release energy during deposition, freezing, and condensation. The result is a decrease in the kinetic energy of the particles of a substance.

5. **A**

A is correct because as a solid changes to a liquid, the particles gain energy and vibrate faster.

B is incorrect because the mass of a substance is conserved between changes of state.

C is incorrect because the identity of the substance remains the same during a change of state.

D is incorrect because gases, not liquids, expand to fill their containers.

Lesson 1 Alternative Assessment

Examples: Examples describe ways that people can measure volume, mass, and density, and give examples of objects that can be measured using each method.

Illustrations: Diagrams show several ways the rooms could be arranged, and show several ways to measure the total interior volume of the home.

Analysis: Explanations tells how to find the mass and weight of an object on Earth and the same object on the moon. Students compare similarities and differences between mass and weight in these two places.

Observations: Students use displacement or a formula to determine the volumes of two similarly sized classroom objects. Descriptions compare the volumes, masses, and densities of the two objects.

Calculations: Students determine the volume of the swimming pool, and the volume of water needed to fill it. They compare the mass, volume, and density of two balls for the pool.

Lesson 2 Alternative Assessment

What Is It?: Descriptions list three things the liquid might be, and how to use the liquid's physical and chemical properties to determine which of the three liquids it is.

Trading Properties: Trading cards identify at least five common physical properties, and they ex-plain and give examples of the properties.

Be the Scientist!: Experiments include a hypothesis and a method for testing one chemical property of an object.

Watch the Reaction: Talks describe the observations and identify the type of property shown.

Advising a Friend: Explanations tell what the friend did, explain why the observation will not identify a chemical property, and suggest what should be done.

Showing Properties: Paragraphs or diagrams describe at least four physical properties of a classroom object.

Lesson 3 Alternative Assessment

Everyday Science: Descriptions explain a way the law of conservation of mass can be applied to daily life.

Chemical Observations: Journal entries describe a common chemical change, how the change occurs, and the results of the change.

Did It Disappear?: Skits explain what happened to the drink left on the counter, and use the law of conservation of mass in the explanation.

Charting Chemical Change: Maps show a backyard with an iron chair, a fire pit filled with wood, and a copper downspout. Maps identify three chemical changes that could occur.

Picturing Change: Posters or collages show both physical and chemical changes occurring to matter. Labels describe the changes.

Water, Water Everywhere: Dialogues explain how water changes from one state to the other.

Cooking up Change: Presentations identify physical and chemical changes that can occur while cooking.

Going in Reverse: Essays list examples of physical and chemical changes that can be reversed, and discuss which type of change seems to be more common.

Lesson 4 Alternative Assessment

Salty and Sweet: Students identify the elements that make up table salt and table sugar.

Make a Model: Models explain the difference between an element and a compound.

Table It: Tables or graphic organizers identify different types of mixtures and their characteristics. Two food products that are examples of each type of mixture are listed.

Parts of Your Entrée: Plans tell how to separate the mixture. Plans identify each part of the mixture and the physical properties that allow separation.

Recipe for Success: Recipe cards include three or more ingredients and a suggestion for how they can be combined to make the food.

Metals, Nonmetals, and Metalloids: Description wheels are drawn for each type of element. On the spokes of the wheel are examples of metals, nonmetals, and metalloids and the properties of each.

Concept Mapping: Concept maps use all seven terms.

Shake Before Use: Students identify three foods that are suspensions and explain why the substance needs to be shaken before use.

Lesson 5 Alternative Assessment

Examples: Presentations follow a substance in more than one state and show how the molecules move in each state.

Illustrations: Posters show molecules in solids, liquids, and gases. Each image indicates how close the molecules are to one another and how they move.

Analysis: Analyses explain whether dough behaves more like a solid or a liquid, and why.

Observations: Critiques compare and critique two animations that show how water looks as a liquid, a solid, and a gas.

Details: Skits describe what is seen out a window at different times during an ice storm. Skits explain how water changes shape and volume at different times of the day.

Lesson 6 Alternative Assessment

Vocabulary: Answers for each term include a personal definition, a dictionary definition, and two sentences that use the term correctly.

Details: The reason why the mass of a substance does not change when its state changes is accurately described. Students should explain that when a substance changes state, the size and number of particles do not change. Only the movement of the particles and the distance between them change in the three states of matter.

Illustrations: Illustrations should accurately portray a liquid changing to a solid and a solid changing to a liquid. The former picture should be labeled *freezing* and the latter picture should be labeled *melting*.

Examples: Answers should show an understanding of the processes of evaporation and condensation. Everyday examples of evaporation include water in a puddle turning to water vapor and alcohol-based hand sanitizer evaporating from one's hands. Examples of condensation include water droplets forming on the bathroom mirror after someone takes a shower and dewdrops on grass in the morning.

Comparisons: Comparisons should accurately describe the movement of particles in solids, liquids, and gases. The particles in a solid vibrate back and forth in the same position and are held tightly together by forces of attraction. The particles in a liquid are constantly sliding around and tumbling over each other as they move. The particles in a gas are far apart and move around at high speeds. The particles do not interact much. When the temperature of a substance rises or drops, its particles speed up or slow down, respectively. If the particles speed up or slow down enough, the substance will change state.

Performance-Based Assessment

See Unit 1, Lesson 1

1. Sample answer: The luster and appearance of the pennies resemble the luster and appearance of the copper sample. The luster and appearance of the nickels, and dimes resemble the luster and appearance of the zinc sample.
2. Sample answer: I think the penny is made of copper because it is the same color as the copper sample. I think the nickel and dime are mostly made of zinc because they are almost the same color as the zinc sample.
3–9. Values in student tables may vary slightly. The calculated densities of copper and zinc should be close to 8.96 g/cm^3 and 7.13 g/cm^3, respectively. The calculated density of the pennies should be the same as the density of copper.
10. All of the coins are primarily made of copper. The densities of the coins are close to the densities of the pure metals that primarily make up the coins.

Unit Review
Vocabulary
1. T See Unit 1, Lesson 1
2. T See Unit 1, Lesson 4
3. F See Unit 1, Lesson 6
4. T See Unit 1, Lesson 5
5. T See Unit 1, Lesson 3

Key Concepts
6. B 10. C 14. C
7. C 11. D 15. C
8. D 12. C 16. D
9. D 13. A 17. A

6. **B** See Unit 1, Lesson 2

A is incorrect because density does not describe a substance's ability to form new substances.

B is correct because flammability describes how likely a substance is to react with oxygen gas in a combustion reaction.

C is incorrect because changing a substance's shape will not change its identity.

D is incorrect because a dissolving a substance does not affect its identity.

7. **C** See Unit 1, Lesson 5

A is incorrect because particles that make up solids do vibrate in place.

B is incorrect because particles in a liquid slide past one another.

C is correct because the particles that make up matter are constantly in motion.

D is incorrect because the particles that make up matter are constantly in motion whether they are solids, liquids, or gases.

8. **D** See Unit 1, Lesson 6

A is incorrect because placing the balloon in the freezer will decrease the kinetic energy of the particles, resulting in an increased attraction between particles.

B is incorrect because placing the balloon in the freezer will decrease the kinetic energy of the particles.

C is incorrect because decreased kinetic energy increases the attraction between particles.

D is correct because freezing the balloon decreases the kinetic energy of the particles inside it. The decreased kinetic energy increases the attraction between the particles.

9. **D See Unit 1, Lesson 5**

A is incorrect because solids, not liquids, have a definite shape and volume.

B is incorrect because gases, not liquids, have neither a definite shape nor a definite volume.

C is incorrect because matter cannot have a definite shape and a variable volume.

D is correct because liquids have a definite volume and take the shape of their container.

10. **C See Unit 1, Lesson 4**

A is incorrect because pure substances cannot be broken down through physical means.

B is incorrect because water molecules cannot be combined with other substances through physical means. Changing the makeup of a water molecule would require a chemical change.

C is correct because the particles that make up pure substances are identical throughout the substance.

D is incorrect because this explains why water is considered a compound, but it does not explain why it is a pure substance.

11. **D See Unit 1, Lesson 5**

A is incorrect because a gas has particles that are further apart than the particles of a liquid.

B is incorrect because a gas has particles that are further apart than the particles of a solid.

C is incorrect because the particles of a solid are not further apart than the particles of a liquid.

D is correct because a change of phase from a liquid to a gas means the particles will be move fur-ther apart from one another.

12. **C See Unit 1, Lesson 3**

A is incorrect because the law of conservation of mass applies to chemical changes as well.

B is incorrect because the law of conservation of mass applies to physical changes as well.

C is correct because mass cannot be created or destroyed by physical or chemical changes.

D is incorrect because the law of conservation of mass applies to matter undergoing physical and chemical changes.

13. **A See Unit 1, Lesson 2**

A is correct because changing from ice to water is a change of state, which is a physical change.

B is incorrect because changing state does not change the identity of a substance.

C is incorrect because a new substance will not form. Melting is a physical change.

D is incorrect because the chemical identity of the ice will not change as it turns to liquid.

14. **C See Unit 1, Lesson 6**

A is incorrect because 0° C is the freezing and melting point of water.

B is incorrect because 32° C is too low of a temperature for water to boil.

C is correct because 100° C is the temperature at which a substance changes from a liquid to a gas.

D is incorrect because 212° Fahrenheit, not Celsius, is the boiling point of water.

15. **C See Unit 1, Lesson 1**

A is incorrect because 40 mL is the total volume of water when the rock is submerged.

B is incorrect because 14 mL is the mark that the rock reaches in the graduated cylinder.

C is correct because 5 mL is the amount of water displaced by the rock. 40 mL – 35 mL = 5 mL or 5 cm^3.

D is incorrect because 35 mL is the amount of water in the graduated cylinder before the rock is added.

16. **D See Unit 1, Lesson 1**

A is incorrect because gravity cannot be measured with a balance.

B is incorrect because a scale is used to measure weight.

C is incorrect because density is the mass to the volume of an object divided by its volume.

D is correct because a balance is used to measure mass.

17. A See Unit 1, Lesson 4

A is correct because a compound is made up of more than one type of atom joined together.

B is incorrect because an element contains only one type of atom.

C is incorrect because mixtures contain multiple substances that are not chemically bonded to each other. Box C shows just one substance.

D is incorrect because each particle of the substance in box C is made up of three atoms joined together.

Critical Thinking
18. See Unit 1, Lesson 4

- Indicates that solutions are homogeneous and one substance dissolves in another.
- Indicates that suspensions are heterogeneous and some particles settle out of solution
- Indicates that colloids are heterogeneous with particles unevenly distributed.
- Gives example of solution, such as salt water or brass
- Gives example of suspension, such as homemade salad dressing or sandy water
- Gives example of colloid, such as whipped cream, gelatin dessert, milk, or mayonnaise.

19. See Unit 1, Lesson 6

- Indicates that the size of particles does not change

- Indicates that the number of particles does not change
- Indicates that the average speed of particles increases

20. See Unit 1, Lesson 3

- identifies physical change does not cause a change in chemical properties
- identifies chemical change produces new substances
- provides three of the following examples of physical changes: freezing, cutting, dissolving, melting, changes in state, changes in texture, changes in appearance.
- provides three of the following signs that a chemical change has taken place: changes in color or odor, production of heat, fizzing and foaming, sound or light being given off
- indicates that temperature can increase or decrease the rate of chemical change

Connect Essential Questions
21. See Unit 1, Lesson 1 and Lesson 2

- Uses correct formula for density: D = m/v
- calculates density of sample as 19.3 g/cm^3
- Identifies the sample as gold
- Lists three of the following: conductivity, specific heat, magnetism, melting point, malleability, ductility, color, luster, texture, state of matter at room temperature

Unit Test A
Key Concepts

1. D	6. A	11. D
2. C	7. C	12. A
3. C	8. C	13. C
4. B	9. A	14. B
5. D	10. B	15. C

1. D

A is incorrect because a scale measures weight. She has already measured the mass of the seashell. She does not need to find its weight.

B is incorrect because she needs to make another measurement before she can perform a calculation.

C is incorrect because density depends on mass and volume. Temperature is not part of the calculation.

D is correct because she has already measured mass and she needs to measure volume. She can measure the volume of an irregularly-shaped object by measuring the amount of liquid it displaces in a graduated cylinder.

2. C

A is incorrect because all types of matter can be described by both physical and chemical properties.

B is incorrect because matter has physical properties that are independent of chemical changes.

C is correct because physical properties can be observed without attempting to change the composition of matter, whereas chemical properties cannot.

Answer Key

   D is incorrect because matter in all states has physical properties.

3. C

   A is incorrect because the particles themselves are not changing.

   B is incorrect because a decrease in volume may accompany a chemical change or a physical change.

   C is correct because the particles are getting closer together, but they do not otherwise change. This model represents a change in state, which is a physical change.

   D is incorrect because the particles are getting closer together, but they are not changing into other substances.

4. B

   A is incorrect because the mass of the ice cream is not increasing during melting.

   B is correct because mass is conserved during a change of state.

   C is incorrect because mass does not increase during freezing.

   D is incorrect because mass is not lost when a substance freezes.

5. D

   A is incorrect because she would need to use a balance to measure mass.

   B is incorrect because the rock is being placed in water. She would need to use a ruler to measure length.

   C is incorrect because the weight of the rock depends on the force of gravity pulling down on it and not the amount of water it displaces.

   D is correct because the volume of water displaced by an object that sinks is equal to the volume of the object.

6. A

   A is correct because each atom that makes up an element or molecule that makes up a compound is identical.

   B is incorrect because only elements are listed on the periodic table.

   C is incorrect because each atom that makes up an element or molecule that makes up a compound is identical.

   D is incorrect because neither elements nor compounds can be broken down by physical changes.

7. C

   A is incorrect because you would not see tiny particles floating in a solution.

   B is incorrect because you would probably see tiny particles in both liquids if they were suspensions.

   C is correct because a solution would be homogeneous, like A, while a suspension would show tiny particles, like B.

   D is incorrect because you see tiny particles floating in B, not A, so A is not a suspension, and B is not a solution.

8. C

   A is incorrect because burning is a chemical change; new substances are formed.

   B is incorrect because a rusting is a chemical change; new substances are formed.

   C is correct because flattening is a physical change. It does not change the chemical makeup of a substance.

   D is incorrect because decomposing is a chemical change; new substances are formed.

9. A

   A is correct because physical properties, such as magnetism, can be used to separate parts of a mixture.

   B is incorrect because the properties of a compound differ from the properties of its elements.

   C is incorrect because the parts of a mixture keep their own properties.

   D is incorrect because, although this statement is true, it does not reflect what is shown in the diagram.

10. B

   A is incorrect because chopping a tree changes its physical shape, not its chemical composition.

   B is correct because cooking a steak changes the chemical composition of the steak.

   C is incorrect because heating a cup of tea changes the tea's temperature, not its chemical composition.

Answer Key

D is incorrect because drying clothes in a dryer changes their moisture, not their chemical composition.

11. D

A is incorrect because the law is not limited to only these types of changes.

B is incorrect because mass is also conserved in chemical changes.

C is incorrect because mass is also conserved in physical changes.

D is correct because mass is conserved in all physical and chemical changes.

12. A

A is correct because liquids and gases take the shape of their containers, but solids do not.

B is incorrect because liquids also have a definite volume.

C is incorrect because the particles in solids vibrate in place. They do not stop moving.

D is incorrect because the particles in all matter are always in motion.

13. C

A is incorrect because the energy absorbed during this change of state acts to break the attractions between particles of matter.

B is incorrect because energy must be absorbed for a liquid to change into a gas. Temperature would not decrease as a result.

C is correct because there is no temperature change while a substance is changing from one state to another.

D is incorrect because temperature is related to the average speed of the substance's particles. Temperature does not decrease and increase during this change of state.

14. B

A is incorrect because the water in the puddle is not heated enough to boil.

B is correct because the water changes state from a liquid to a gas at temperatures that are lower than the boiling point of water.

C is incorrect because condensation occurs when a gas changes into a liquid. The water in the puddle is already a liquid.

D is incorrect because water particles in the puddle do not undergo a chemical change to become another substance.

15. C

A is incorrect because color is a physical property rather than a chemical property.

B is incorrect because the freezing point of a substance is a physical property.

C is correct because the reactivity of a substance is a chemical property. Reactivity can be observed only by altering the nature of the substance.

D is incorrect because volume is a physical property.

Critical Thinking

16.

- description of motion of particles in solids, liquids, and gases (e.g., *The particles in a solid are tightly packed together and vibrate in place. The particles in a liquid can slide past one another. The particles in a gas can move freely*; etc.)

- description of a process for modeling particle motion in solids, liquids, and gases (e.g., *I could model a solid by tilting the box so that all the marbles are tightly packed on one end of the box. I would then move the box back and forth slightly so that the marbles move in place. I could model a liquid by slightly tilting the box so that all the marbles are at one end of the box. I would then move the box so that the marbles still touch, but they slide by each other. I could model a gas by shaking the box rapidly so that the marbles move quickly in all directions and are not by each other*; etc.)

Extended Response

17.

- sublimation

- description of sublimation (e.g., *The solid changes directly to a gas without first passing through the liquid state*; etc.)

- comparison of particle motion in dry ice and carbon dioxide gas (e.g., *The particles in dry ice are close together and vibrate in place, whereas the particles in carbon dioxide gas move rapidly in all directions*; etc.)

- comparison of original masses and explanation of answer (e.g., *The mass of dry ice equals the mass of carbon dioxide gas formed from it. This is true because mass is conserved in all changes of state*; etc.)

Unit Test B
Key Concepts
1. D   6. C   11. B
2. B   7. A   12. B
3. B   8. A   13. B
4. A   9. D   14. A
5. A   10. D   15. C

1. D

A is incorrect because if the objects were of equal mass, the pans would be balanced.

B is incorrect because the two cubes are identical in size so they have the same volume.

C is incorrect because the pans would be balanced if the cubes had the same weight.

D is correct because the cubes have equal volumes, but the pans are not balanced. Therefore, they have different masses. Density is mass divided by volume, so they must have different densities.

2. B

A is incorrect because the tarnishing of silverware is a chemical property, not a physical property.

B is correct because texture is a physical property.

C is incorrect because the effect of acid rain on automobiles is a chemical property, not a physical property.

D is incorrect because the combustion of gasoline in a car engine is a chemical property, not a physical property.

3. B

A is incorrect because the particles themselves are not changing so the mass stays the same.

B is correct because the appearance of the particles, rather than their chemical composition, is changing. This is a physical change.

C is incorrect because only the arrangement of the particles is changing.

D is incorrect because the particles are getting closer together. The particles themselves are not changing in a way that would make them more reactive.

4. A

A is correct because the number of water particles is the same throughout the process, so mass is conserved.

B is incorrect because the mass of a substance does not change during a change of state, even if its volume does change.

C is incorrect because no water particles are lost during changes of state.

D is incorrect because the change in density that results from a change in volume does not affect the mass.

5. A

A is correct because the water level rose from 35 mL to 40 mL. Therefore, the rock displaced 5 mL of water. The volume of water displaced by an object that sinks is equal to the volume of the object.

B is incorrect because the rock displaced 5 mL of water, not 10 mL.

C is incorrect because 35 mL is the initial amount of water in the cylinder, not the displaced amount of water.

D is incorrect because 40 mL is the final level of water in the cylinder, not the displaced amount of water.

6. C

A is incorrect because both elements and compounds are pure substances.

B is incorrect because neither elements nor compounds can be broken down by physical changes.

C is correct because an element is composed of identical atoms, and a compound is composed of identical groups of atoms, or molecules.

D is incorrect because an element is composed of identical atoms, and a compound is composed of identical groups of atoms, or molecules.

7. A

A is correct because a light beam would travel through water as it does in container A.

B is incorrect because gelatin is a colloid and so would block some of the light.

C is incorrect because apple juice is a solution and so would block some of the light.

D is incorrect because mayonnaise is a colloid and so would block some of the light.

8. A

A is correct because changing from liquid to gas is a state change. This process is a physical change.

B is incorrect because synthesis of new compounds is a chemical change, not a physical change.

C is incorrect because decomposition of compounds is a chemical change, not a physical change.

D is incorrect because decomposition of matter is a chemical change, not a physical change.

9. D

A is incorrect because a solution has one substance dissolved in another.

B is incorrect because the pile is not a suspension.

C is incorrect because the magnet could not separate a compound easily.

D is correct because the pile is a simple mixture of iron and sulfur and one part can be clearly distinguished from another.

10. D

A is incorrect because burning is not necessarily a function of the type of bread.

B is incorrect because the size of the bread would not speed up the chemical change.

C is incorrect because decreasing a lower temperature would most likely slow down, not speed up, a chemical change.

D is correct because chemical changes are affected by temperature.

11. B

A is incorrect because chemical changes observe the law of conservation of mass, so mass will not be gained during the reaction.

B is correct because chemical changes observe the law of conservation of mass, so the mass will not change.

C is incorrect because although gas is less dense than a solid or a liquid, mass is conserved during a chemical change and is not lost.

D is incorrect because although gas can change in volume, mass is conserved during a chemical change and is not lost.

12. B

A is incorrect because only the particles in gases move as far apart as possible.

B is correct because the particles in all matter are always in motion.

C is incorrect because the shape of a solid does not depend on its container.

D is incorrect because only solids do not take the shape of the container.

13. B

A is incorrect because matter is made up of particles in all three states. A sample of matter might melt, but the particles within it do not.

B is correct because temperature is related to the average speed of the particles. As temperature increases, the average speed of the particles in a sample increases.

C is incorrect because the particles themselves remain the same despite the change of state.

D is incorrect because heating does not decrease the weight of particles of matter.

14. A

A is correct because fog is made up of drops of water that form through condensation.

B is incorrect because drops of water that fall form rain or snow. Fog is a cloud near the ground.

C is incorrect because liquid water on the ground does evaporate, but it forms water vapor in the air. Fog is made up of liquid water rather than water vapor.

D is incorrect because air particles do not change into water molecules to form fog.

15. C

A is incorrect because there is more water in the second beaker, so their masses are not the same.

B is incorrect because the second beaker has more water. The water in the second beaker is heavier than that in the first, so their weights are not the same.

C is correct because samples of pure water will have the same density.

D is incorrect because the water in the second beaker takes up more space than that in the first, so their volumes are not the same.

Critical Thinking
16.

- glass window is solid; raindrops are liquids; air is gas

- comparison of attractions between particles in glass window, raindrops, and air (e.g., *The attraction between the particles in order of greatest to least is the glass window, the raindrop, and the air. The glass is a solid, so the attraction between the particles is strong enough to lock the particles in one place. The raindrop is a liquid, so the attraction between the particles is strong enough to hold the particles together, but not strong enough to prevent them from sliding past one another. The air is a gas, so there is very little attraction between the particles and they can move freely*; etc.)

Extended Response
17.

- 0 °C

- comparison of particle motion in solid, liquid, and gas (e.g., *The particles in the ice vibrate in place. The particles in liquid water move more than in ice, so that they slide by each other. The particles in a gas move quickly in all directions*; etc.)

- comparison of particle attraction in solid, liquid, and gas (e.g., *The attraction between particles in a solid is the greatest. The attraction between particles in a liquid is less than in a solid. The attraction between particles in a gas is the least*; etc.)

- comparison of mass in ice and water vapor (e.g., *The mass remains the same because mass is conserved during a change of state. So the mass of the ice, liquid water, and water vapor is a constant 20 g*; etc.)

Unit 2 Energy
Unit Pretest
1. A   5. B   9. D
2. B   6. B   10. C
3. B   7. A
4. C   8. A

1. A

A is correct because using alternative energy resources can reduce the use of fossil fuels.

B is incorrect because alternative energy resources are used instead of fossil fuels in order to reduce the use of fossil fuels.

C is incorrect because a nonrenewable resource is a resource that is used faster than it can be replaced.

D is incorrect because all energy resources have pros and cons with regard to the environment.

2. B

A is incorrect because nuclear energy is stored in the attractions between particles in the centers of atoms. Batteries store a different kind of energy.

B is correct because a battery stores chemical energy.

C is incorrect. Although a battery can be used to produce electrical energy, a battery does not store electrical energy.

D is incorrect because a battery can lead to mechanical energy, but mechanical energy is not stored in a battery.

3. B

A is incorrect because the diagram represents a liquid. Particles in liquids move faster than particles in solids but not faster than particles in gases.

B is correct because the diagram represents a gas. Particles in gases move faster than particles in solids and liquids.

C is incorrect because the diagram represents a solid. Particles in solids move more slowly than particles in liquids and gases.

Answer Key

D is incorrect because diagram A represents a liquid, diagram B represents a gas, and diagram C represents a solid. Particles in gases move faster than particles in solids and liquids.

4. C

A is incorrect because kinetic energy and potential energy can be transformed into one another.

B is incorrect because any object can have kinetic energy, potential energy, or both.

C is correct because the sum of all energy does not change, although it can be converted into other forms.

D is incorrect because energy is never created or destroyed; it is only transformed.

5. B

A is incorrect because people typically drill when they're trying to obtain fossil fuels; using fossil fuels typically involves burning them, which can cause air pollution.

B is correct because obtaining fossil fuels by drilling into land or the ocean floor can destroy habitats and pollute water and soil.

C is incorrect because transporting fossil fuels might cause spills that can cause pollution that harms the environment, but it doesn't typically involve drilling.

D is incorrect because converting fossil fuels to usable forms can produce harmful byproducts, but it doesn't typically involve drilling.

6. B

A is incorrect because the gravitational potential energy continues to increase as the ball rises.

B is correct because this is the farthest point above the ground.

C is incorrect because some of the gravitational potential energy has been converted into kinetic energy.

D is incorrect because most of the gravitational potential energy has been converted into kinetic energy.

7. A

A is correct because the freezing point of water is 273 K. The freezing point of corn oil is about 259 K. Thus, 273 K − 259 K = 14 K lower.

B is incorrect because the freezing point of water is 273 K. The freezing point of corn oil is about 259 K. Thus, 273 K − 259 K = 14 K lower, not 47 K.

C is incorrect because the freezing point of water is 273 K. The freezing point of corn oil is about 259 K. Thus, 273 K − 259 K = 14 K lower, not 114 K.

D is incorrect because the freezing point of water is 273 K. The freezing point of corn oil is 259 K. Thus, 273 K − 259 K = 14 K lower, not 159 K.

8. A

A is correct because thermal energy is the total kinetic energy of all particles in a substance.

B is incorrect because this statement describes the temperature of a substance, not its thermal energy.

C is incorrect because thermal energy is the total kinetic energy of all particles in a substance. It does not include potential energy.

D is incorrect because thermal energy is the total kinetic energy of all particles in a substance. It does not include potential energy.

9. D

A is incorrect because melting takes place when a solid gains energy.

B is incorrect because evaporation takes place when a liquid gains energy.

C is incorrect because solidification takes place when a liquid loses energy.

D is correct because the gas particles lose energy, and forces of attraction cause them to condense into a liquid.

10. C

A is incorrect because radiation is the process of transferring energy through electromagnetic waves.

B is incorrect because insulation is a process that inhibits the transfer of energy between objects.

C is correct because convection is the transfer of energy by the movement of currents within a fluid.

D is incorrect because conduction is the transfer of energy from one particle to another without the movement of matter.

Lesson 1 Quiz
1. A   4. D
2. D   5. C
3. B

1. A

A is correct because thermal energy is transferred in the form of heat from an object at a higher temperature to an object at a lower temperature.

B is incorrect because chemical energy is stored in the bonds between the atoms of molecules.

C is incorrect because electrical energy is due to the attraction or repulsion of charged particles.

D is incorrect because mechanical energy involves potential and kinetic energy.

2. D

A is incorrect because the law of conservation of energy applies to closed systems.

B is incorrect because a force cannot create or destroy energy, but can only transform energy into a different form or transfer it from one object to another.

C is incorrect because heating and cooling objects transfer energy and does not destroy or create energy.

D is correct because the law of conservation of energy states that energy cannot be created or destroyed but can be transformed or transferred.

3. B

A is incorrect because kinetic energy is the energy of motion.

B is correct because potential energy is the energy stored within the slingshot as a result of the position of the band.

C is incorrect because no charges are passing through the band.

D is incorrect because there is no light released from the slingshot.

4. D

A is incorrect because nuclear energy is released when atoms come together or split apart.

B is incorrect because chemical energy is stored in the bonds between the atoms of molecules.

C is incorrect because electrical energy is the energy due to the attraction or repulsion of charged particles.

D is correct because x-rays are a form of electromagnetic energy.

5. C

A is incorrect because a bicyclist coasting down hill is a conversion of gravitational potential energy into kinetic energy.

B is incorrect because a car engine moving a car is a conversion of chemical potential energy into kinetic energy.

C is correct because the kinetic energy of the car is converted into gravitational potential energy as the height of the car increases.

D is incorrect because a thrown ball after the top part of it motion is a conversion of gravitational potential energy into kinetic energy.

Lesson 2 Quiz
1. B   4. A
2. D   5. B
3. B

1. B

A is incorrect because 82°C + 8°C = 90 °C, but the boiling point of water is 100 °C.

B is correct because 82°C + 18°C = 100°C, which is the boiling point of water.

C is incorrect because 82°C + 130°C = 212°C. A temperature of 212 °F is the boiling point of water on the Fahrenheit scale, but a temperature of 100 °C is the boiling point of water on the Celsius scale.

D is incorrect because 82°C + 191°C = 273°C. A temperature of 273 K is the freezing point of water on the Kelvin scale, but a temperature of 100°C is the boiling point of water on the Celsius scale.

2. D

A is incorrect because the particles are not ordered

(liquid). Particles in a solid are close and ordered.

B is incorrect because the particles are too spread out. Particles in a solid are close and ordered.

C is incorrect because the particles are too spread out (gas). Particles in a solid are close and ordered.

D is correct because the particles are close and ordered as in a solid.

3. **B**

A is incorrect because an increase in temperature means the particles of a substance have more kinetic energy and thus are free to move farther apart.

B is correct because an increase in temperature causes the particles of a substance to move faster on average.

C is incorrect because an increase in temperature means the particles of a substance have more average kinetic energy, not less.

D is incorrect because only particles of a solid vibrate and are lose together, not a liquid such as soup.

4. **A**

A is correct because 385 K – 12 K = 373 K. This is the correct boiling point of water on the Kelvin scale.

B is incorrect because 385 K – 73 K = 312 K, which is not the boiling point of water.

C is incorrect because 385 K – 112 K = 273 K, which is not the boiling point of water.

D is incorrect because 385 K – 173 K = 212 K, which is not the boiling point of water.

5. **B**

A is incorrect because all of the particles that make up matter are constantly in motion.

B is correct because according to the kinetic theory of matter, particles of matter are constantly in motion.

C is incorrect because the particles of a liquid, not a solid, slide past each other.

D is incorrect because the particles that make up a solid do not move from place to place.

Lesson 3 Quiz
1. C   4. C
2. B   5. A
3. B

1. **C**

A is incorrect because radiation is the transfer of energy by electromagnetic waves.

B is incorrect because insulation inhibits the transfer of energy, and the difference in temperature is a result of energy transfer.

C is correct because energy is transferred in the room through currents of moving air.

D is incorrect because conduction takes place when energy is transferred from one particle to another without the movement of matter. Warm air in a room is carried by air currents.

2. **B**

A is incorrect because a boiling liquid becomes a gas as heat increases, not decreases.

B is correct because as heat increases, a solid becomes liquid, the liquid boils, and finally the boiling liquid becomes a gas.

C is incorrect because a solid becomes a liquid as heat increases, not decreases.

D is incorrect because a solid becomes a liquid, and a liquid becomes a gas, as heat increases, not decreases.

3. **B**

A is incorrect because temperature is a measure of average kinetic energy, and heat is the energy transferred between objects at different temperatures.

B is correct because heat is a transfer of energy between objects at different temperatures, which causes the temperature of both objects to change.

C is incorrect because an increase in temperature is caused by heat, or the transfer of energy between objects.

D is incorrect because temperature is a measurement, but heat energy that is transferred between objects.

4. **C**

A is incorrect because the cubes are not touching, so there is no energy transferred and thus the temperature of both cubes is the same.

B is incorrect because the temperature of both cubes is the same.

C is correct because cube A is larger and has more mass than cube B. Thermal energy is a total amount of energy.

D is incorrect because cube B is smaller and has less mass than cube A.

5. **A**

A is correct because both of these units measure the transfer of energy between objects at different temperatures.

B is incorrect because a degree Celsius is a measure of temperature, not heat.

C is incorrect because the Celsius scale is used to describe temperature rather than heat.

D is incorrect because these units are both used to measure temperature, not heat.

Lesson 4 Quiz

1. C   4. B
2. B   5. A
3. C

1. **C**

A is incorrect because wind and water are forces of weathering.

B is incorrect because lava hardens to form rock, not fossil fuels.

C is correct because heat and pressure help in the process of decomposition of layers of sediments that become fossil fuels over millions of years.

D is incorrect because acids in rainwater can carve out caves over time, but not form fossil fuels.

2. **B**

A is incorrect because coal is a type of fossil fuel. An alternative energy sources is a resource that can be used in place of fossil fuels.

B is correct because biomass is living or recently dead organic material materials that can be used as an energy resource.

C is incorrect because petroleum is a type of fossil fuel. An alternative energy sources is a resource that can be used in place of fossil fuels.

D is incorrect because natural gas is a type of fossil fuel. An alternative energy sources is a resource that can be used in place of fossil fuels.

3. **C**

A is incorrect because wind plants use wind energy, which is a renewable resource.

B is incorrect because solar plants use solar energy, which is a renewable resource.

C is correct because nuclear plants use the element uranium, which is a nonrenewable resource.

D is incorrect because hydroelectric plants use water, which is a renewable resource.

4. **B**

A is incorrect because rocks cannot store the energy produced by the sun.

B is correct because green plants absorb energy from the sun and store it as chemical energy.

C is incorrect because ocean water itself does not store energy from the sun, but some algae in the ocean do absorb energy from the sun and store it as chemical energy.

D is incorrect because although animals rely on energy from the sun, they obtain it in their bodies by eating plants or by eating organisms that eat plants.

5. **A**

A is correct because water is a renewable energy source, but flooding caused by building dams can destroy habitats.

B is incorrect because a nuclear power plant, not a hydroelectric plant, produces waste that is difficult to store.

C is incorrect because a geothermal plant, not a hydroelectric plant, can only be built near hot springs or volcanoes.

D is incorrect because a plant that burns biomass, not a hydroelectric plant, releases carbon dioxide.

Lesson 1 Alternative Assessment

Energy Journal: Journal entries should describe how energy behaves in a closed system.

Making a Transformation: Skits should identify an instance when one form of energy transforms to another form of energy.

Energy Observations: Observations should note examples of energy in the classroom, and include a diagram depicting the room. Diagrams should have labels and describe at least three examples of energy in the room.

Where Did the Energy Go? Sketches should show energy transformations and for all of the initial energy. Forms of energy should be labeled.

Puzzling Terms: Crossword puzzles should contain at least four terms about energy transformation and include an answer key.

Make a Model: Students should use a pendulum to explain how mechanical energy works during the pendulum's movement.

Experimenting with Energy: Experiments should describe the law of conservation of energy, include a hypothesis, and the method to test it.

Oral Presentation: Oral presentation should explain how kinetic energy and potential energy differ, and include some of the many forms of kinetic and potential energy. Demo should show the differences between some of these forms of energy.

Lesson 2 Alternative Assessment

Vocabulary: Writing defines each vocabulary word and explains what each word has to do with temperature.

Illustrations: Poster shows a Fahrenheit, Celsius, and Kelvin thermometer. Thermometers are labeled and show the temperature at which water freezes.

Analysis: Analysis tells how people should prepare for an outdoor event if the weather is 35 degrees C, F, or K.

Observations: Observations are recorded in a notebook, and note the temperatures of ice, cool water, and warm water. Notes describe how particles behave at each temperature.

Details: Presentations describe the Fahrenheit, Celsius, and Kelvin temperature scales, and give details about freezing and boiling points on each temperature scale.

Lesson 3 Alternative Assessment

Vocabulary: Definition of *radiation* is correct. Journal entry compares and contrasts the meanings of words that are related to *radiation*.

Examples: Posters identifies three examples of conductors and three examples of insulators, and de-scribes the qualities that make them good conductors or insulators.

Illustrations: Cartoons show several methods of thermal energy transfer. Captions identify the methods of transfer and explain how the methods are related.

Analysis: Explanations describe why the beverage got cooler and how the thermal energy was transferred.

Details: Presentations describe convection in ocean water, where the water is moving during convection, why it is moving, how water temperature is measured, and how to measure to show thermal energy transfer.

Lesson 4 Alternative Assessment

Be the Teacher: The quiz should cover the important facts from the lesson about renewable and nonrenewable resources, fossil fuels, and alternative energy sources.

Collage: Student collage should include at least six energy resources and should identify whether the resources are renewable or nonrenewable.

Pro/Con Grid: Grid should provide relevant information about the advantages and disadvantages of the energy resources featured in the lesson.

Skit: Student skit should include discussion of the key ideas from the lesson, including advantages and disadvantages of various energy sources.

Commercial: Commercials should explain how the product uses an alternative energy source and how it compares to products that use other types of energy resources.

Poem: Poem should include an analysis of how at least one aspect of human energy use affects the environment.

Invention: Design should focus on how the machine using one or more alternative energy sources. An explanation of how the machine would affect the environment should be included.

Puzzle Time: Puzzle should include key terms from the lesson. Clues should show that student understand key ideas.

Web Site: Site should include at least three pages and should include a glossary of terms, examples of various energy sources, and images.

Performance-Based Assessment

See Unit 2, Lesson 3

1. Answers may vary. Sample answer: I think the temperature of the ice water will rise as the ice melts.

3. 0°C, or 32°F

6. The trend line remained flat until the ice cubes melted, and then the line curved up. This is because the temperature remained constant while the ice cubes melted, but then the temperature began to rise after the last ice cube had melted.

7. Answers may vary. Sample answer: The energy that was added during the change of state went to breaking the attractions between the particles that made up the ice rather than to changing the temperature of the ice water.

8. Answers may vary. Sample answer: Once all the attractions between the particles in the ice had been broken and the change of state from ice to liquid was complete, the heat energy from the hot plate went towards raising the temperature of the water.

Unit Review

Vocabulary

1. F See Unit 2, Lesson 4
2. T See Unit 2, Lesson 1
3. F See Unit 2, Lesson 4
4. F See Unit 2, Lesson 2
5. T See Unit 2, Lesson 3

Key Concepts

6. D 9. C 12. C
7. A 10. B 13. C
8. D 11. D

6. D See Unit 2, Lesson 3

A is incorrect because heat is the energy transferred from an object of higher temperature to an object of lower temperature. These objects have the same temperature.

B is incorrect because calories are a way to measure heat.

C is incorrect because if two objects have the same temperature, the object with more particles will have more overall kinetic energy.

D is correct because if two objects have the same temperature, the object with more particles will have more overall kinetic energy.

7. A See Unit 2, Lesson 4

A is correct because there are many types of alternative energy sources that can be used in place of fossil fuels.

B is incorrect because solar energy is energy from the sun, and is just one type of alternative energy resource.

C is incorrect because nuclear energy is energy from splitting nuclei, and is just one type of alternative energy resource.

D is incorrect because biomass energy is energy from plant material, and is just one type of alternative energy resource.

8. D See Unit 2, Lesson 1

A is incorrect because electromagnetic energy is transmitted through space in the form of electromagnetic waves.

B is incorrect because mechanical energy is the sum of an object's kinetic and potential energy.

C is incorrect because sound energy is the energy of vibration of particles in a medium such as air or water.

D is correct because chemical energy is stored in chemical bonds that hold the atoms of substances, such as those found in food, together.

9. C See Unit 2, Lesson 1

A is incorrect because as the spring stretches downward, some energy is also in the form of kinetic energy.

B is incorrect because the spring has both potential energy and kinetic energy.

C is correct because as the spring stretches downward energy is changing forms from potential to kinetic energy.

D is incorrect because kinetic energy is the energy of movement and potential energy is the stored energy that an object has due to its position. The spring in Position 1 has both kinetic and potential energy.

10. **B See Unit 2, Lesson 3**

A is incorrect because conduction is the transfer of energy as heat through direct contact.

B is correct because convection requires the movement of a liquid or gas for heat transfer.

C is incorrect because emission is not the term that describes the transfer of energy as heat by the movement of a liquid or gas.

D is incorrect because radiation is the transfer of heat by electromagnetic waves.

11. **D See Unit 2, Lesson 3**

A is incorrect because heat is the energy transferred from an object at a higher temperature to an object at a lower temperature.

B is incorrect because temperature is a measure of the average kinetic energy of the particles in an object.

C is incorrect because thermal energy is the total kinetic energy of all particles in a substance.

D is correct because a calorie is the amount of energy needed to raise the temperature of 1 gram of water by 1 degree Celsius.

12. **C See Unit 2, Lesson 2**

A is incorrect because a barometer measures air pressure, not temperature.

B is incorrect because a scale is used to measure weight, not temperature.

C is correct because a thermometer is used to measure temperature.

D is incorrect because a balance is used to measure mass.

13. **C See Unit 2, Lesson 3**

A is incorrect because wood is a good insulator but not a good conductor of energy as heat.

B is incorrect because most metals are good conductors but not good insulators of energy.

C is correct because a conductor transmits energy well.

D is incorrect because an insulator does not transmit energy well.

Critical Thinking

14. **See Unit 2, Lesson 1**

- states that energy cannot be created or destroyed
- explains that energy changes form
- provides two examples of energy transformations (e.g., *roller coaster moving down a hill and potential energy changing to kinetic energy*; etc.)

15. **See Unit 2, Lesson 2**

- uses the scale to find equivalent Celsius temperature at approximately 16° C
- uses the scale to find equivalent Kelvin temperature at approximately 288 K
- explains that the average kinetic energy of particles decreases with decreased temperature

Connect Essential Questions

16. **See Unit 2, Lesson 2 and Lesson 3**

- identifies that direct contact transfers energy through conduction
- explains that adding energy in the form of heat may cause the ice to melt
- explains that adding energy in the form of heat may cause the water to evaporate
- recognizes that the particles in the ice do not move much
- recognizes that the particles in the water move more than those in the ice, but less than those in the air
- recognizes that the particles in the air move faster than those in the ice or water

Unit Test A
Key Concepts

1. B   5. B   9. D
2. B   6. B   10. D
3. C   7. D   11. C
4. B   8. D   12. C

1. **B**

A is incorrect because temperature is a measure of average kinetic energy, so it does not depend on mass.

B is correct because a thermometer reading is a

measure of the temperature of a substance. The two thermometers both show readings of 60°C.

C is incorrect because temperature is a measure of average kinetic energy, and thermal energy is the total kinetic energy.

D is incorrect because temperature is an instantaneous reading, and the thermometers indicate the same temperatures.

2. **B**

A is incorrect because energy is not lost but is converted to other forms.

B is correct because the potential energy of the ball is converted to kinetic energy, as the ball rolls down the ramp.

C is incorrect because energy cannot be destroyed but can only change in form.

D is incorrect because kinetic energy affects the speed of the ball, not potential energy. Also, the ball speeds up as it rolls down the ramp, it does not slow down.

3. **C**

A is incorrect because kinetic energy increases as mass increases.

B is incorrect because kinetic energy depends on another factor in addition to speed.

C is correct because the object with the greatest mass has the greatest kinetic energy when the objects are moving at the same speed.

D is incorrect because kinetic energy is a result of motion, not the number of tires.

4. **B**

A is incorrect because 1 calorie is the amount of energy needed to raise the temperature of 1 gram of water by 1 degree Celsius; therefore, 10 calories are needed to raise the temperature of 1 gram of water by 10 degrees Celsius.

B is correct because 1 calorie is the amount of energy needed to raise the temperature of 1 gram of water by 1 degree Celsius; therefore, 10 calories are needed to raise the temperature of 1 gram of water by 10 degrees Celsius.

C is incorrect because 1 calorie is the amount of energy needed to raise the temperature of 1 gram of water by 1 degree Celsius; therefore, 10 calories are needed to raise the temperature of 1 gram of water by 10 degrees Celsius.

D is incorrect because 1 calorie is the amount of energy needed to raise the temperature of 1 gram of water by 1 degree Celsius; therefore, 10 calories are needed to raise the temperature of 1 gram of water by 10 degrees Celsius.

5. **B**

A is incorrect because biomass provides only a small portion of the energy resources used in the United States.

B is correct because most of the energy used in the United States comes from fossil fuels.

C is incorrect because solar energy provides only a small portion of the energy resources in the United States.

D is incorrect because wind energy provides only a small portion of the energy resources in the United States.

6. **B**

A is incorrect because heat causes a transfer in thermal energy, not in the size of molecules.

B is correct because heat is the transfer of thermal energy, which causes the ice cube's molecules to move faster.

C is incorrect because heat is the transfer of energy into the ice cube as it melts, not the removal of thermal energy.

D is incorrect because a change in state does not cause the bonds between atoms in a molecule to break.

7. **D**

A is incorrect because particles vibrate in all matter, whether warm or cold.

B is incorrect because particles move less freely if a material is colder.

C is incorrect because particles speed up as they are heated.

D is correct because particles have less energy in a colder material such as the colder air in the math classroom.

8. D

A is incorrect because a balance would be used to measure mass.

B is incorrect because the student would need to measure mass and volume to calculate density.

C is incorrect because a graduated cylinder might be best for measuring volume.

D is correct because the tool shown is a thermometer, and it is used to measure temperature.

9. D

A is incorrect because electrical energy and sound energy are forms of energy, but magnetic energy is not.

B is incorrect because thermal energy is a form of energy, but electronic energy and magnetic energy are not.

C is incorrect because mechanical energy and nuclear energy are forms of energy, but geothermic energy is not.

D is correct because electromagnetic energy, mechanical energy, and sound energy are all forms of energy.

10. D

A is incorrect because some renewable resources come from plants, but others do not.

B is incorrect because the supply of renewable resources is not unlimited, although renewable resources can be replaced much more quickly than nonrenewable resources can be replaced.

C is incorrect because every energy resource has some cost associated with it.

D is correct because unlike nonrenewable resources, renewable resources can be replaced at the same rate at which they are used.

11. C

A is incorrect because the melting point is not the same as the boiling point.

B is incorrect because 32 degrees is the freezing point on the Fahrenheit scale, but these answer choices are in degrees Celsius.

C is correct because the freezing/melting point of water is 0 °C and the boiling point of water is 100 °C.

D is incorrect because 273 is associated with events on the Kelvin scale.

12. C

A is incorrect because biomass energy production contributes more to air pollution than it does to the destruction of habitats.

B is incorrect because biomass energy production contributes more to air pollution than it does to erosion.

C is correct because contributing to air pollution is the biggest environmental drawback of using biomass as an energy source.

D is incorrect because biomass energy production does not generate hazardous waste.

Critical Thinking
13.
- freezing point of water: 273 K
- boiling point of water: 373 K
- 198 K below the freezing point of water
- 298 K below the boiling point of water

Extended Response
14.
- direct contact, or conduction
- description of what will happen to the spoons (e.g., *The silver and stainless steel spoons will feel hot, the plastic spoon will feel warm, and the wood spoon temperature will not have changed*; etc.)
- insulators: plastic and wood; conductors: stainless steel and silver
- explanation of how radiation could affect the system (e.g., *Radiation could come in the form of a heat source at some distance away, such as a fire under the bowl or sunlight falling on the spoons and bowl*; etc.)

Unit Test B
Key Concepts
1. B	5. A	9. D
2. A	6. A	10. A
3. B	7. C	11. C
4. D	8. B	12. B

1. B

A is incorrect because although the two liquids have the same temperature, the one with more particles has a higher total kinetic energy.

Answer Key

B is correct because although the two liquids have the same temperature, the one with more particles has a higher total kinetic energy. Therefore, the beaker with 1,000 mL has more thermal energy than the beaker with 100 mL.

C is incorrect because the relative amounts of thermal energy can be determined with the information provided. Although the two liquids have the same temperature, the one with more particles has a higher total kinetic energy. Therefore, the beaker with 1,000 mL has more thermal energy than the beaker with 100 mL.

D is incorrect because although the two liquids have the same temperature, the one with more particles has a higher total kinetic energy. Therefore, the beaker with 1,000 mL has more thermal energy than the beaker with 100 mL.

2. **A**

A is correct because as the baseball is rising, its kinetic energy is being converted to potential energy.

B is incorrect because when the baseball is falling, its potential energy is being converted to kinetic energy.

C is incorrect because there would be no energy conversion while the baseball sits on the ground.

D is incorrect because no energy conversion is taking place while the baseball is still.

3. **B**

A is incorrect because larger objects have more kinetic energy than smaller objects if they are traveling at the same speed.

B is correct because the motorcycle has the least mass and must travel the fastest to have the same kinetic energy as the other vehicles.

C is incorrect because the delivery van has the most mass, so it must be traveling the slowest of all the other vehicles in order to have the same kinetic energy as the others.

D is incorrect because kinetic energy is related to the motion at a particular time, not the place and time of the start of motion.

4. **D**

A is incorrect because 1 calorie is the amount of energy needed to raise the temperature of 1 gram of water by 1 degree Celsius; therefore, 100 calories are needed to raise the temperature of 10 grams of water by 10 degrees Celsius.

B is incorrect because 1 calorie is the amount of energy needed to raise the temperature of 1 gram of water by 1 degree Celsius; therefore, 100 calories are needed to raise the temperature of 10 grams of water by 10 degrees Celsius.

C is incorrect because 1 calorie is the amount of energy needed to raise the temperature of 1 gram of water by 1 degree Celsius; therefore, 100 calories are needed to raise the temperature of 10 grams of water by 10 degrees Celsius.

D is correct because 1 calorie is the amount of energy needed to raise the temperature of 1 gram of water by 1 degree Celsius; therefore, 100 calories are needed to raise the temperature of 10 grams of water by 10 degrees Celsius.

5. **A**

A is correct because drilling can destroy habitats and pollute water and soil.

B is incorrect because hydroelectric power is more likely to disrupt fish migration.

C is incorrect because burning fossil fuels will release greenhouse gases into the air.

D is incorrect because using nuclear power produces radioactive wastes.

6. **A**

A is correct because heat is the transfer of thermal energy, which causes a substance's molecules to move faster. Therefore, the loss of heat causes a substance's molecules to move slower.

B is incorrect because heat is the transfer of energy into an

Answer Key

ice cube as it melts, not the removal of thermal energy from liquid water as it freezes.

C   is incorrect because heat causes a transfer in thermal energy, not in the size of molecules.

D   is incorrect because a change in state does not cause the bonds between atoms in a molecule to break.

7. **C**

A   is incorrect because the air becomes warmer, not cooler.

B   is incorrect because the particles are farther apart in warm air than in cool air.

C   is correct because particles in a substance have more energy if the substance is warm. More energy means the particles move more.

D   is incorrect because the air is warmer, but it did not change to a different state of matter.

8. **B**

A   is incorrect because the top of the liquid is between 70 °F and 80 °F. The temperature is 75 °F.

B   is correct because the top of the liquid is between 70 °F and 80 °F. The temperature is 75 °F.

C   is incorrect because the top of the liquid is between 70 °F and 80 °F. The temperature is 75 °F.

D   is incorrect because the top of the liquid is between 70 °F and 80 °F. The temperature is 75 °F.

9. **D**

A   is incorrect because chemical energy is stored in the chemical bonds of a substance's atoms.

B   is incorrect because electromagnetic energy is transmitted through space in the form of electromagnetic waves.

C   is incorrect because mechanical energy is the sum of an object's kinetic and potential energy.

D   is correct because thermal energy, or heat, is the energy an object has due to the motion of its particles.

10. **A**

A   is correct because human activities can alter the status of a resource.

B   is incorrect because all energy resources have some negative impact on the environment.

C   is incorrect because businesses can purchase and control both renewable and nonrenewable resources.

D   is incorrect because a resource is described as renewable according to its supply and usage.

11. **C**

A   is incorrect because changing from a liquid to a solid does not change the freezing point of a substance.

B   is incorrect because the temperature of the liquid water is higher, not lower, than the temperature of the ice.

C   is correct because temperature is a measure of the average kinetic energy of the particles of a substance. Because the temperature of the ice is lower than the temperature of the liquid water, the average kinetic energy of the ice particles is lower.

D   is incorrect because the temperature of the liquid water is higher. This means the particles have more energy and move faster, not slower.

12. **B**

A   is incorrect because nuclear energy generally destroys fewer, not more, habitats.

B   is correct because nuclear energy generally produces less air pollution than burning fossil fuels does.

C   is incorrect because nuclear energy generally creates less, not more, land erosion.

D   is incorrect because nuclear energy produces waste that remains harmful on Earth for many years.

Critical Thinking
13.
- 87 °F − 32 °F = 55 °F cooler
- 212 °F − 87 °F = 125 °F warmer

Extended Response
14.
- comparison of temperature (e.g., *The temperature of the water in beaker 1 is greater than in the other two beakers, which have the same temperature*; etc.)

- comparison of average motion (e.g., *The average motion of the particles in beakers 2 and 3 are the same. The average motion of the particles in beaker 1 is greater than that of the other beakers. This is because temperature is a measure of the average kinetic energy of the particles of the substance. Beaker 1 is warmer than either beakers 2 or 3, so its average particle motion is greater*; etc.)

- comparison of thermal energy (e.g., *The thermal energy of beaker 1 is the greatest, and the thermal energy of beaker 3 is the least. Beaker 2 has twice as much thermal energy as beaker 3 because it is at the same temperature but contains twice as much mass. Beaker 1 has more thermal energy than beaker 2 does because they have the same mass, but beaker 1 has a higher temperature*; etc.)

- comparison of heat flow (e.g., *Beaker 1 is at a higher temperature than beaker 2, so thermal energy will flow from beaker 1 to beaker 2 until they are at the same temperature. Beaker 2 is at the same temperature as beaker 3, so they are in equilibrium. An equal amount of thermal energy will flow between them in such a way that their temperatures remain the same*; etc.)

Unit 3 Atoms and the Periodic Table

Unit Pretest

1. B   5. C   9. A
2. D   6. C   10. B
3. B   7. D
4. A   8. A

1. B

A is incorrect because there are two electrons in the innermost energy level, not the outermost energy level.

B is correct because the outermost energy level in the Bohr model contains five electrons.

C is incorrect because the atom contains a total of seven electrons, but they are not all valence electrons.

D is incorrect because the atom only contains seven electrons, five of which are valence electrons.

2. D

A is incorrect because the atomic mass depends on the atomic number and the number of neutrons, which requires additional information.

B is incorrect because the number of neutrons cannot be determined from the atomic number or the place of an element on the periodic table.

C is incorrect because mass number depends on the atomic number and the number of neutrons, which requires additional information.

D is correct because atoms are arranged on the periodic table in order of increasing atomic number.

3. B

A is incorrect because a chemical bond does not necessarily involve metals.

B is correct because a chemical bond is a force of attraction between atoms.

C is incorrect because chemical bonds result in chemical changes. In a chemical change, the substances that come together change to form a new substance.

D is incorrect because a chemical bond involves atoms, which are too small to be physically pulled apart.

4. A

A is correct because the strong attractions between ions in the crystal lattice result in high melting and boiling points.

B is incorrect because ionic compounds are usually brittle.

C is incorrect because ionic compounds are often solids at room temperature.

D is incorrect because ionic compounds usually conduct electricity when dissolved in water.

5. C

A is incorrect because all the particles are labeled incorrectly.

B is incorrect because the neutron and nucleus are labeled incorrectly.

C is correct because all the particles are labeled correctly.

D is incorrect because the proton and neutron are labeled incorrectly.

6. C

A is incorrect because each element has a unique atomic number.

B is incorrect because elements are not arranged on the periodic table based on their chemical symbols.

C is correct because elements are arranged in groups based on shared chemical properties.

D is incorrect because elements in a group do not have the same average atomic mass. Average atomic mass typically increases from the top of a column to the bottom.

7. D

A is incorrect because an ion forms when an atom gains or loses electrons.

B is incorrect because a solid ionic compound forms a crystal lattice.

C is incorrect because an attraction between positively charged metal ions and free electrons is a metallic bond.

D is correct because a group of atoms joined by covalent bonds makes up a molecule.

8. A

A is correct because an atom is the smallest particle of an element that retains its properties.

B is incorrect because a proton is a positively charged particle associated with elements, but protons of different elements are the same.

C is incorrect because a molecule does not have the properties of the elements that make it up.

D is incorrect because electrons are not unique for different elements.

9. A

A is correct because during metallic bonding, metal atoms give up their outer electrons to form positive ions in a sea of electrons.

B is incorrect because the diagram does not show ionic bonding in which electrons are transferred between different kinds of atoms.

C is incorrect because the atoms are not sharing electrons as happens during covalent bonding.

D is incorrect because the electrons come from the atoms that are shown as ions.

10. B

A is incorrect because metals lie to the left of the zigzag line.

B is correct because most of the elements that border the zigzag line are neither metals nor nonmetals, but have properties of both.

C is incorrect because nonmetals lie to the right of the zigzag line.

D is incorrect because transition metals are in the center of the table.

Lesson 1 Quiz
1. B   4. A
2. A   5. A
3. D

1. B

A is incorrect because 85 is the number of protons in the atom.

B is correct because the atom's mass number minus its atomic number is equal to the number of neutrons in the atom.

C is incorrect because 210 is the mass number, which is equal to the total number of protons and neutrons in the atom.

D is incorrect because adding the atomic number and the mass number of an atom does not yield the number of neutrons.

2. A

A is correct because all matter is composed of atoms.

B is incorrect because mass number is a quantity used to describe a particular atom, not an entire object.

C is incorrect because the chair is made up of wood and the balloon is made up of rubber and filled with air. They are composed of different substances.

D is incorrect because the chair and the balloon are likely made up of a different number of atoms.

3. **D**

A is incorrect because in 1887, J. J. Thomson performed an experiment that showed that there are small, negatively-charged particles within the atom. These particles are now called *electrons*.

B is incorrect because in 1887, J. J. Thomson performed an experiment that showed that there are small, negatively-charged particles within the atom. These particles are now called *electrons*.

C is incorrect because in 1887, J. J. Thomson performed an experiment that showed that there are small, negatively-charged particles within the atom. These particles are now called *electrons*.

D is correct because in 1887, J. J. Thomson performed an experiment that showed that there are small, negatively-charged particles within the atom. These particles are now called *electrons*.

4. **A**

A is correct because an atom is the smallest particle of an element that has the chemical properties of that element.

B is incorrect because a proton from a calcium atom does not have the properties associated with calcium.

C is incorrect because an electron from a calcium atom does not represent the properties of calcium.

D is incorrect because a molecule that contains calcium would have different properties than calcium.

5. **A**

A is correct because A points to the center of the atom, called the nucleus, which contains the protons and neutrons.

B is incorrect because B points to a neutron.

C is incorrect because C points to a proton.

D is incorrect because D points to the electron cloud.

Lesson 2 Quiz

1. C   4. C
2. D   5. C
3. C

1. **C**

A is incorrect. A group is a column of boxes, not a single box.

B is incorrect. A period is a row of boxes, not a single box.

C is correct because each box contains the name of an element and information about that element.

D is incorrect. A compound is made of two or more elements.

2. **D**

A is incorrect because while some metals are in the first group, not all metals are in the first group.

B is incorrect because while some metals are in the last period, not all metals are in the last period, and some of the elements in the last period are not metals.

C is incorrect because all of the elements in this column are gases.

D is correct because all the metals are located to the left of the zigzag line.

3. **C**

A is incorrect because liquid is a state of matter, not a kind of element.

B is incorrect because metals are conductors, not partial conductors.

C is correct because metalloids have only some of the properties of metals, such as conductivity. Some metalloids are used to make semiconductors in electronic devices.

D is incorrect because nonmetals are nonconductors, not partial conductors.

4. **C**

A is incorrect because Rb, rubidium, has an average atomic mass of 85.47.

B is incorrect because Sc, scandium, has an average atomic mass of 44.96.

C is correct because Sr, strontium, has an average atomic mass of 87.62.

D is incorrect because Y, yttrium, has an average atomic mass of 88.91.

5. **C**

A is incorrect because chemical symbols are not used to arrange elements in order.

B is incorrect because although early attempts at the periodic table used atomic mass, the

Answer Key

modern periodic table uses a different criterion.

C is correct because the atomic number increases by one with each successive square going from left to right, across each period.

D is incorrect because there is no relationship between the name of the element and its location on the periodic table.

Lesson 3 Quiz
1. A 4. C
2. D 5. A
3. B

1. A

A is correct because spheres can be used to show the relationship among atoms, but not the details within individual atoms.

B is incorrect because charged particles would need to be located within the spheres and could not be represented in the model.

C is incorrect because energy levels are not solid spheres and therefore cannot be represented by solid spheres.

D is incorrect because spheres cannot be used to distinguish the nucleus from the rest of the atom.

2. D

A is incorrect because matter is conserved during a chemical change.

B is incorrect because the identity of an atom remains the same during a chemical change.

C is incorrect because atoms do not become larger or smaller during a chemical change.

D is correct because atoms are neither created nor destroyed. Instead, bonds between atoms are broken, atoms are rearranged, and new bonds are formed during chemical changes.

3. B

A is incorrect because protons and neutrons, not electrons, are found in the central nucleus.

B is correct because the Bohr model illustrates electrons in at different energy levels around the nucleus.

C is incorrect because protons and neutrons are located in the nucleus, whereas electrons are located elsewhere.

D is incorrect because scientists did not begin to visualize electrons as clouds around the nucleus until after Bohr developed his model.

4. C

A is incorrect because the boiling point relative to water does not indicate whether or not an element will form a chemical bond.

B is incorrect because the relationship between protons and neutrons is not what determines if an element will form a chemical bond.

C is correct because sodium has one valence electron and can become stable by giving up

that electron and forming a chemical bond.

D is incorrect because atomic number is not the property that determines the reactivity of an element.

5. A

A is correct because all elements in the same group (column) have the same number of valence electrons. Oxygen and sulfur are in the same group.

B is incorrect because chlorine has 7 valence electrons rather than 6.

C is incorrect because bromine is not in the same group as sulfur.

D is incorrect because even though phosphorus is in the same period, it is in a different group than sulfur. Phosphorus has one less valence electron than sulfur.

Lesson 4 Quiz
1. A 4. A
2. D 5. D
3. C

1. A

A is correct because the positively charged metal ions are in fixed positions in the metal, whereas the electrons are free to move around them. Moving electric charges can establish an electric current.

B is incorrect because this attraction establishes a metallic bond rather than an electric current.

C is incorrect because metals consist of metal ions and free electrons. The bond is not an equal sharing as in some covalent bonds.

D is incorrect because this property explains why metals do not carry a charge. It does not explain why metals are good conductors of electricity.

2. **D**

A is incorrect because ionic compounds are generally solids at room temperature.

B is incorrect because ionic compounds are soluble in water and will readily dissolve.

C is incorrect because ionic compounds have high melting and boiling points because of the strong attractions between ions in the crystal lattice.

D is correct because although ionic solids are poor conductors of electricity, solutions of ionic compounds in water are good conductors of electrical current.

3. **C**

A is incorrect because a sodium atom does not become an ion by losing a proton.

B is incorrect because gaining a proton will change the identity of an element.

C is correct because a sodium atom loses one valance electron to become an ion.

D is incorrect because gaining one electron would produce a negatively charged ion.

4. **A**

A is correct because covalent bonds generally form between nonmetals, such as carbon and fluorine.

B is incorrect because ionic bonds generally form between a positively-charged metal ion and a negatively-charged nonmetal ion. Sodium forms a positively-charged metal ion and chlorine forms a negatively-charged nonmetal ion.

C is incorrect because potassium and calcium are both metals. Atoms of these elements will not become stable by sharing electrons.

D is incorrect because magnesium is a metal and oxygen is a nonmetal. These elements are more likely to form an ionic bond.

5. **D**

A is incorrect because an ion forms when an atom gains or loses electrons. According to the diagram, neither carbon nor oxygen gain or lose electrons to form carbon dioxide.

B is incorrect because a mixture can be separated by physical means, but a chemical change occurred to form carbon dioxide. The atoms are bonded together.

C is incorrect because carbon and oxygen are elements. The diagram shows how the elements combine to form a unit described by a different name.

D is correct because the diagram shows that carbon and oxygen atoms are held together by covalent bonds. A group of atoms held together by covalent bonds is called a molecule.

Lesson 1 Alternative Assessment

Atomic Cooking: Recipes include the parts of an atom (protons, neutrons, and electrons), their charges, and their locations. The element's atomic number and mass number are also correctly identified.

Atom's Next Top Model: The atomic number and mass number of the chosen element is correctly represented in the model. The model's nucleus and electron cloud are clearly identified. The protons and neutrons that are located in the nucleus and the electrons that are located in the electron cloud are also clearly identified.

Atoms Flipping Out: For the chosen element, flipbooks show the nucleus and the electron cloud of an atom. Protons and neutrons, which are located inside the nucleus, are correct in number and are clearly shown. The correct number of electrons is shown traveling in the electron cloud, not in orbits similar to planets traveling around the sun.

Atoms Acting Out! Plays present all parts of the current atomic theory.

Read the Small Print: Newspaper interviews correctly summarize Democritus' or Aristotle's view on atoms and

Answer Key

describe how their view contributed to the current atomic theory.

That's a Funny Theory: Cartoon or comic strips include all parts of the current atomic theory, as well as some early ideas that contributed to it.

Lesson 2 Alternative Assessment

Get Organized: Tables classify elements as metals, nonmetals, or metalloids, and list the name and chemical symbol of each element.

All in the Family: Displays identify and illustrate six properties shared by the elements making up one chemical family.

Periodic Interview: Newspaper articles, newscasts, or radio broadcasts air an interview with Dmitri Mendeleev about his contribution to the current arrangement of the elements. Works discuss his arrangement, how he came to his conclusion, whether there are any problems with the arrangement, and explain the current arrangement.

Body of Elements: Posters show three elements that make up the human body and explain how each is used by the body. Labels include the element's name, symbol, atomic number, average atomic mass, and in what percentage the element is found.

What's in a Name?: Tables show three elements whose symbols are not made up from the letters of the element's name. Tables explain why the chemical symbol was selected and include each element's name, symbol, atomic number, and average atomic mass.

Illustrating Mass Differences: Drawings show the isotopes that play a role in determining an element's average atomic mass. Protons, neutrons, atomic numbers, and masses of each element are labeled.

Lesson 3 Alternative Assessment

Vocabulary: Student stories about chemical bonding are told from the point of view of an atom or electron and correctly use the words *chemical bond, chemical change, molecule, valence electrons,* and *periodic table.*

Examples: Students should have taken several photos that show chemical changes. Their descriptions for each photo should correctly explain the chemical change that occurred and what happed at the molecular level.

Illustrations: Students should draw Bohr models for several elements. Models should have the electrons in the correct energy levels. Students should correctly note how many valence electrons each atom has, predict how likely each atom is to form bonds with other atoms, and explain how they can use the periodic table to determine how the atom will interact with other atoms.

Analysis: Student explanations to the class should correctly state why electrons are sometimes shared while other atoms are transferred from one atom to another.

Observations: Student summaries should include their observations of the chemical change that took place, as well as their research on the chemical reaction that occurred. Summaries should also correctly identify and present models of the reactants and products.

Calculations: Students should calculate the correct number of valence electrons in several elements in Groups 3 through 12. Students should record their results and should note whether or not they noticed any patterns.

Lesson 4 Alternative Assessment

Model: Student models should clearly show how each of the three types of bonds is formed. Models should be clearly labeled and neat.

Story: Stories should explain how each of the three types of bonds is formed. Stories should be well organized and logical.

Crossword Puzzle: Crossword puzzle should include all of the key terms from this lesson and any other relevant or appropriate words related to this lesson. Clues given for the terms should show an understanding of the terms.

Song: Songs should include a description of how each of the three types of bond is formed and should describe the properties of each of the three types.

Flipbook: Flipbooks should include an explanation of the different ways that each of the three types of bonds form.

Guessing Game: The clues that the student writes should demonstrate his or her understanding of the distinct properties that result from each of the three types of bonds.

Performance-Based Assessment

See Unit 3, Lesson 3

1. Answers may vary. Accept reasonable answers.
2. Answers may vary. Sample answer: The steel wool looks like metallic cotton. The color is gray or dark silver.
5. Answers may vary. Sample answer: The starting temperature is 24°C.
8. Answers may vary. Sample answer: The ending temperature is 30°C.
9. The parts of the steel wool that were soaked in vinegar appear rusty. The color changed from gray to orange.
10. Answers may vary. Sample answer: Rust appeared, and the temperature of the steel wool rose by 6°C.
11. Answers may vary. Sample answer: Iron has two valence electrons. Oxygen has six valence electrons. Because these elements do not have full outer energy levels, they are reactive. The iron and oxygen reacted with one another, forming rust.
12. The steel wool would rust very quickly. Chlorine is more reactive than oxygen because it has seven valence electrons (it only needs to gain 1 electron to become full).
13. The steel wool and neon would not react. Neon is a noble gas (a full outer energy level) and is therefore non-reactive.
14. Answers may vary. Sample answer: Knowing the position of the element in the periodic table makes it easier to predict the physical and chemical properties of the element. That is because elements in the same vertical column in the periodic table are part of the same group or family and thus share similar properties.

Unit Review

Vocabulary

1. **ionic bond** See Unit 3, Lesson 4
2. **chemical bond** See Unit 3, Lesson 3
3. **atom** See Unit 3, Lesson 1
4. **covalent bond** See Unit 3, Lesson 4
5. **electron** See Unit 3, Lesson 1

Key Concepts

6. B   9. A   12. B
7. C   10. B   13. D
8. D   11. A

6. **B** See Unit 3, Lesson 4

A is incorrect because metals generally have high melting points.

B is correct because the electrons in a metal can move freely, so most metals are good conductors of electric current.

C is incorrect because metals are good thermal conductors.

D is incorrect because metals are malleable, and can bend without breaking.

7. **C** See Unit 3, Lesson 1

A is incorrect because Element A has 10 protons.

B is incorrect because 20 is the number of protons and neutrons for Element A.

C is correct because the atomic number of an element equals the number of protons in an atom of that element.

D is incorrect because 19 is the mass number of Element B.

8. **D** See Unit 3, Lesson 1

A is incorrect because Thomson's model showed electrons dispersed evenly throughout an atom.

B is incorrect because Dalton did not mention subatomic particles in his theory.

C is incorrect because Rutherford's model did not show electrons arranged in energy levels.

D is correct because Bohr's model showed electrons orbiting a nucleus in rings representing energy levels.

9. **A** See Unit 3, Lesson 3

A is correct because the number of electrons equals the number of protons in a neutral atom.

B is incorrect because there could be more or fewer neutrons than electrons in a neutral atom.

Answer Key

C is incorrect because there is only one nucleus per atom.

D is incorrect because there are varying numbers of electrons in each energy level of an atom.

10. B See Unit 3, Lesson 2

A is incorrect because the average atomic mass is given at the bottom of the square, and the atomic number is given at the top of the square.

B is correct because the atomic number is given at the top of the square, and the average atomic mass is given at the bottom.

C is incorrect because the top number is the atomic number, and the bottom number is the average atomic mass.

D is incorrect because the last number is the average atomic mass, not the proton number.

11. A See Unit 3, Lesson 4

A is correct because an ion is a charged particle that forms when an atom gains or loses one or more electrons.

B is incorrect because an atom forms an ion when it gains or loses an electron, not a proton.

C is incorrect because a change in neutron number in an atom can form an isotope, but an atom must gain or lose an electron to form an ion.

D is incorrect because an atom cannot gain or lose a nucleus and remain an atom. An ion is formed by an atom gaining or losing electrons.

12. B See Unit 3, Lesson 2

A is incorrect because the vertical columns of the periodic table are called groups, not periods, and energy levels are not used to describe the periodic table. The horizontal rows are called periods.

B is correct because the vertical columns are called groups, and the horizontal rows are called periods.

C is incorrect because atomic numbers are listed for each element in the individual squares of the periodic table. The vertical columns of the periodic table are called groups, and the horizontal rows are called periods.

D is incorrect because atomic numbers are listed for each element in the individual squares of the periodic table. The vertical columns of the periodic table are called groups, and the horizontal rows are called periods.

13. D See Unit 3, Lesson 1

A is incorrect because electrons travel in Rutherford's model and the current model.

B is incorrect because electrons are fixed only in Thomson's model.

C is incorrect because electrons travel in orbits in Rutherford's model.

D is correct because the current model of the atom shows electrons within a region around the nucleus called an electron cloud.

Critical Thinking

14. See Unit 3, Lesson 2

- identifies three of the following properties of metals: shiny, ductile, good electrical and thermal conductors, malleable
- describes metals as being to the left of the zigzag line on the periodic table and nonmetals to the right
- identifies metalloids as having some properties of both metals and nonmetals

15. See Unit 3, Lesson 3

- recognizes that 2 atoms of fluorine will form a covalent bond
- draws 2 fluorine atoms sharing electrons to fill their outermost energy levels
- expresses that atoms bond to fill their outermost energy levels

Connect Essential Questions

16. See Unit 3, Lesson 1 and Lesson 3

- drawings show electrons as dots arranged in rings around the nucleus
- labels the electrons in the outermost rings of the atom as valence electrons
- identifies element by the number of protons in the nucleus
- expresses that Bohr models show the number of electrons that are available for bonding

Unit Test A
Key Concepts
1. B   5. A   9. D
2. B   6. D   10. C
3. A   7. C   11. A
4. B   8. A   12. B

1. B

A is incorrect because the atoms do not change into other kinds of atoms.

B is correct because bonds are broken, atoms are rearranged, and new bonds are formed during this change.

C is incorrect because no atoms are destroyed during a chemical change.

D is incorrect because the new substance that is formed has different properties than the substances from which the new substance was formed.

2. B

A is incorrect because the atomic number is the number of protons.

B is correct because subtracting the atomic number from the mass number gives you the number of neutrons in an atom.

C is incorrect because the mass number is equal to the number of protons and neutrons in an atom.

D is incorrect because adding the mass number and the atomic number does not yield useful information.

3. A

A is correct because the sodium atom loses one valance electron to become an ion.

B is incorrect because the atom is not emitting particles from its nucleus.

C is incorrect because the number of neutrons is not changing.

D is incorrect because the atomic number of the atom stays the same, so the atom remains the same element.

4. B

A is incorrect because if nitrogen were in the first group, it would have just one valence electron. This would imply that carbon has no valence electrons based on the information given, and that is not true.

B is correct because elements in the same group have the same number of valence electrons. Nitrogen and carbon must be in different groups.

C is incorrect because elements that have different numbers of valence electrons are in different groups.

D is incorrect because elements that have different numbers of valence electrons are in different groups.

5. A

A is correct because Rb and Cs are in the same vertical column, or group.

B is incorrect because Ca and Ti are in the same row, or period, but not the same group.

C is incorrect because Li and Be are in the same row, or period, but not the same group.

D is incorrect because Na and Mg are in the same row, or period, but not the same group.

6. D

A is incorrect because free electrons move around positive ions in a metal.

B is incorrect because bonds are not formed by destroying electrons.

C is incorrect because electrons are transferred during ionic bonding.

D is correct because atoms share electrons to become stable during covalent bonding.

7. C

A is incorrect because the negative charges would have balanced the positive charge, so this discovery would not have contradicted Thomson's model.

B is incorrect because a discovery about the structure of the atom required a revision.

C is correct because a small positive center, later known as the nucleus, did not resemble plum pudding.

D is incorrect because Thomson's model did not identify exact locations, so this discovery did not oppose the model.

8. A

A is correct because the elements along the zigzag line have some properties of metals and some properties of nonmetals.

B is incorrect because the highlighted elements are not gases.

C is incorrect because these elements can react with other elements.

D is incorrect because these elements occur in nature.

9. D

A is incorrect because atoms are the building blocks of all matter, including elements.

B is incorrect because atoms are the building blocks of all matter, including molecules.

C is incorrect because atoms are the building blocks of all matter, including pure substances, which include both elements and molecules.

D is correct because subatomic particles such as protons, electrons, and neutrons are particles of atoms.

10. C

A is incorrect because nonmetals are usually brittle and dull.

B is incorrect because most metals are solids at room temperature.

C is correct because metals can usually be drawn into wires and hammered into sheets.

D is incorrect because metals are described by different properties.

11. A

A is correct because metallic bonds are weak when compared to ionic or covalent bonds.

B is incorrect because the strengths of the different bonds are not all the same.

C is incorrect because covalent bonds are stronger than metallic bonds.

D is incorrect because ionic bonds are stronger than metallic bonds.

12. B

A is incorrect because the Bohr model is not a solid sphere.

B is correct because Bohr models show electrons circling the nucleus much like planets circle the sun.

C is incorrect because Bohr did not arrange the parts of the atom in a row like beads on a string.

D is incorrect because the Bohr model does not show protons, neutrons, and electrons as being joined together.

Critical Thinking
13.
- 19 protons
- explains why potassium has 19 protons (e.g., *Each atom of potassium has 19 protons because the atomic number is the number of protons*; etc.)
- 19 electrons
- explains why potassium has 19 electrons (e.g., Each neutral atom of potassium has 19 electrons because in a neutral atom the number of electrons is equal to the number of protons; etc.)

Extended Response
14.
- protons, electrons, and neutrons
- comparison of charge (e.g., *Protons are positively charged particles. Electrons are negatively charged particles. Neutrons have no charge*; etc.)
- comparison of mass (e.g., *Protons and neutrons have about the same mass. Electrons are much less massive than protons or neutrons*; etc.)
- description of atomic structure (e.g., *Protons and neutrons are located in the central nucleus of an atom, whereas electrons travel around the nucleus*; etc.)

Unit Test B
Key Concepts
1. D   5. D   9. D
2. C   6. A   10. D
3. B   7. C   11. A
4. D   8. D   12. C

1. D

A is incorrect because atoms are not destroyed during chemical changes.

B is incorrect because the new substances formed have different properties than the substances from which the new substances were formed.

C is incorrect because the atoms are simply rearranged during the chemical change.

D is correct because the atoms of water are separated and rearranged to form hydrogen and oxygen gases.

Answer Key

2. C

A is incorrect because 4 is the number of protons or the number of electrons, not the mass number.

B is incorrect because 5 is the number of neutrons, not the mass number.

C is correct because the mass number is the number of protons, 4, plus the number of neutrons, 5.

D is incorrect because 13 is the sum of the protons, electrons, and neutrons, and electrons are not involved in finding the mass number.

3. B

A is incorrect because carbon has four valence electrons. Taking one electron from sodium will not make it stable.

B is correct because chlorine is a nonmetal with seven valence electrons. It needs only one more to become stable.

C is incorrect because metals, such as sodium, generally bond with nonmetals that accept their electrons.

D is incorrect because magnesium is a metal that also needs to give away electrons to become stable.

4. D

A is incorrect because the alkali metals have 1 valence electron.

B is incorrect because elements in the carbon group have 4 valence electrons.

C is incorrect because the halogens have 7 valence electrons.

D is correct because the noble gases have a complete set. For most of the elements in this group, a complete set consists of 8 electrons. For helium, a complete set consists of 2 electrons.

5. D

A is incorrect because Mg, Ca, and Sr have different atomic masses.

B is incorrect because Mg, Ca, and Sr have different numbers of protons and neutrons.

C is incorrect because the atomic number is unique for every element.

D is correct because all of the elements in the same vertical group have the same number of valence electrons.

6. A

A is correct because covalent substances dissolve less readily than ionic substances.

B is incorrect because ionic compounds have higher melting and boiling points.

C is incorrect because ionic compounds, not covalent substances, have a crystalline structure and are brittle.

D is incorrect because ionic compounds generally conduct electric current better once dissolved in water.

7. C

A is incorrect because most of the volume of an atom is still believed to be empty space.

B is incorrect because electrons do have a negative charge, and the current atom model reflects this.

C is correct because the cloud represents the probability of finding an electron in a specific region.

D is incorrect because the model does not suggest that electrons form a single mass.

8. D

A is incorrect because brittleness would not make an element useful for electronics.

B is incorrect because the highlighted elements are not gases.

C is incorrect because these elements have different numbers of valence electrons.

D is correct because the highlighted elements are metalloids. Some metalloids are also semiconductors that can be made to act like metals in only some situations.

9. D

A is incorrect because atoms are the building blocks of all matter, including elements.

B is incorrect because atoms are the building blocks of all matter, including molecules.

C is incorrect because atoms are the building blocks of all matter, including pure substances, which include both elements and molecules.

D is correct because atoms are made up of subatomic particles such as protons, electrons, and neutrons.

10. D

A is incorrect because being shiny is not helpful for cooking food.

B is incorrect because nonmetals are usually brittle.

C is incorrect because ductility is not helpful for cooking food.

D is correct because metals can conduct thermal energy to the food for cooking.

11. A

A is correct because the Bohr model identifies the valence electrons that take part in bonding.

B is incorrect because the Bohr model does not show true masses.

C is incorrect because the Bohr model does not show attractions between particles in the nucleus, but instead focuses on the energy levels of electrons.

D is incorrect because the specific location of an electron cannot be identified, especially using a Bohr model.

12. C

A is incorrect because the electrons remain inside the metal.

B is incorrect because metallic bonds pull oppositely charged particles together, but bonding does not remove charges from either particle.

C is correct because when charges are balanced, there is no net charge on the metal.

D is incorrect because electrons are negatively charged particles no matter where they are found.

Critical Thinking
13.
- comparison of valence electrons in potassium and calcium (e.g., *Potassium is in group 1 of the table, so each atom of potassium has 1 valence electron, whereas calcium is in group 2 of the table, so each atom of calcium has 2 valence electrons*; etc.)
- describe bonding in each element (e.g., *Atoms of both elements need a complete set of valence electrons to become stable. A complete set would consists of 8 electrons. Potassium can become stable by giving away 1 electron whereas calcium can become stable by giving away 2 electrons*; etc.)

Extended Response
14.
- description of atomic number (e.g., *The number of protons in the nucleus of an atom determines its atomic number. Each element has a unique atomic number*; etc.)
- description of mass number (e.g., *The total number of particles in the nucleus of an atom determines the mass number of an atom*; etc.)
- description of mass (e.g., *Protons and neutrons are much more massive than electrons. Therefore, the mass of an atom is concentrated in the nucleus, which is where protons and neutrons are located*; etc.)
- explanation of balanced charges (e.g., *Protons carry a positive charge and electrons carry a negative charge that is equal in magnitude. As a result, the charges cancel out so that an atom has no overall charge*; etc.)

Unit 4 Interactions of Matter

Unit Pretest

1. C   5. A   9. D
2. C   6. B   10. D
3. B   7. D
4. C   8. A

1. C

A is incorrect because a catalyst does not work by decreasing the temperature of the reaction.

B is incorrect because a catalyst does not work by decreasing the surface area of the reactant.

C is correct because a catalyst brings reactant particles together so that they are more likely to react.

D is incorrect because catalysts do not react to form new products.

2. C

A is incorrect because the chemical formula for oxygen is incorrect. Oxygen gas is represented by O_2.

B is incorrect because the chemical formula for magnesium oxide is incorrect. Magnesium oxide is represented by MgO.

Answer Key

C is correct because the numbers of magnesium and oxygen atoms is the same on both sides of the equation, and the chemical formulas are correct.

D is incorrect because there are different numbers of magnesium atoms on the right and left sides of the equation, and because the formula for oxygen is incorrect. Oxygen gas is represented by O_2.

3. B

A is incorrect because an alpha particle does not contain electrons. An alpha particle is made up of two protons and two neutrons.

B is correct because an alpha particle is made up of two protons and two neutrons.

C is incorrect because an alpha particle has a charge of 2+, while a positron has a charge of 1+. An alpha particle is made up of two protons and two neutrons.

D is incorrect because an alpha particle does not contain electrons. An alpha particle is made up of two protons and two neutrons.

4. C

A is incorrect because every atom has a single nucleus.

B is incorrect because each atom of an element, such as hydrogen, has the same number of protons.

C is correct because each isotope has a different number of neutrons and

therefore a different mass number.

D is incorrect because the number of electrons is the same as the number of protons.

5. A

A is correct because butane contains only hydrogen and carbon atoms.

B is incorrect because methanal contains oxygen.

C is incorrect because hydrocarbons do not contain oxygen.

D is incorrect because each O represents an atom of carbon.

6. B

A is incorrect because an exothermic reaction releases energy, and an endothermic reaction absorbs energy.

B is correct because an exothermic reaction releases energy, and an endothermic reaction absorbs energy.

C is incorrect because an exothermic reaction releases energy, and an endothermic reaction absorbs energy; energy is neither created nor destroyed in any reaction.

D is incorrect because an exothermic reaction releases energy, and an endothermic reaction absorbs energy; energy is neither created nor destroyed in any reaction.

7. D

A is incorrect because bubbles do form during many chemical reactions, but bubbles do not constitute a precipitate.

B is incorrect because a color change is involved in many chemical reactions, but a precipitate is different from a color change.

C is incorrect because steam forms when gas condenses into drops of liquid water. The formation of steam is a physical change, whereas a precipitate often forms from a chemical change.

D is correct because a precipitate is a solid that sinks to the bottom when a chemical reaction takes place in a liquid.

8. A

A is correct because mass is converted into energy.

B is incorrect because the total mass of all particles changes after the reaction. If some particles escaped the reactor, their mass would not disappear. The reason that mass is less after the reaction is that some mass is converted into energy.

C is incorrect because particles are not squeezed to become smaller.

D is incorrect because some mass is converted into energy.

9. D

A is incorrect because carbon atoms can make up to four bonds with other atoms.

B is incorrect because the number of valence electrons that carbon has to share does not change.

Answer Key

C is incorrect because the size of each hydrogen atom does not depend on the number of bonds that carbon forms.

D is correct because the number of hydrogen atoms in a molecule depends on the kinds of bonds that carbon forms.

10. D

A is incorrect because a nuclear fission reaction is when the nucleus of a large atom splits into two smaller nuclei.

B is incorrect because atoms do not convert their electrons into energy during a nuclear fusion reaction.

C is incorrect because an atom does not gain electrons from other atoms during a nuclear fusion reaction.

D is correct because during fusion reactions, a small amount of mass is converted to energy when the nuclei of smaller atoms combine.

Lesson 1 Quiz
1. C   4. C
2. A   5. C
3. A

1. C

A is incorrect because bonds do not form in the reactants.

B is incorrect because bonds do not break in the products.

C is correct because bonds between atoms in any reactant break apart, and any bonds within the products form.

D is incorrect because during a chemical reaction bonds break in the reactants, not the products, and bonds form in the products, not the reactants.

2. A

A is correct because there are 12 hydrogen atoms, 6 carbon atoms, and 18 oxygen atoms on the left side of the equation, and there must be the same number on the right side of the equation.

B is incorrect because there are 12 hydrogen atoms, 6 carbon atoms, and 18 oxygen atoms on the left side of the equation, and there must be the same number on the right side of the equation. If there were 2 molecules of glucose, there would be 24 hydrogen atoms, 12 carbon atoms, and 24 oxygen atoms on the right side of the equation.

C is incorrect because there are 12 hydrogen atoms, 6 carbon atoms, and 18 oxygen atoms on the left side of the equation, and there must be the same number on the right side of the equation. If there were 6 molecules of glucose, there would be 72 hydrogen atoms, 36 carbon atoms, and 48 oxygen atoms on the right side of the equation.

D is incorrect because there are 12 hydrogen atoms, 6 carbon atoms, and 18 oxygen atoms on the left side of the equation, and there must be the same number on the right side of the equation. If there were 12 molecules of glucose, there would be 144 hydrogen atoms, 72 carbon atoms, and 84 oxygen atoms on the right side of the equation.

3. A

A is correct because cake batter must absorb energy in order to change into a cake.

B is incorrect because a candle releases energy when it burns. Burning is exothermic.

C is incorrect because a firework releases energy when it explodes in an exothermic reaction.

D is incorrect because smoldering is a form of combustion, and combustion is an exothermic process.

4. C

A is incorrect because changing the temperature does not change the number of reactant particles.

B is incorrect because temperature does not change the size of the particles that make up the reactant

C is correct because raising the temperature of the reactants causes them to move faster and collide into each other more often. This increases the chances of collisions leading to a reaction.

D is incorrect because having more space between particles does not explain the increase in the rate of a chemical reaction. If particles are further apart, they are less likely to interact.

5. C

A is incorrect because the atoms in the reactants rearrange, forming products that differ from the reactants.

B is incorrect because sulfur dioxide and sulfur trioxide are the products, not the reactants, in this chemical change.

C is correct because a chemical change produces new substances with properties different from those of the reactants.

D is incorrect because the number of each type of atom should be the same in the products as in the reactants.

Lesson 2 Quiz
1. D   4. D
2. A   5. D
3. C

1. D

A is incorrect because the formula should indicate the number of each type of atom.

B is incorrect because there is more than 1 atom of hydrogen in the diagram.

C is incorrect because there are more than 3 atoms of hydrogen in the diagram.

B is correct because the diagram shows 3 atoms of carbon and 8 atoms of hydrogen.

2. A

A is correct because polymers are long-chain carbon molecules made of repeating structural units.

B is incorrect because many different kinds of parts make up a car. A polymer has repeating parts.

C is incorrect because a bowling ball is not made of many distinct, repeating parts, whereas a polymer is.

D is incorrect because the fibers of a carpet are not arranged in the same way the parts of a polymer are arranged.

3. C

A is incorrect because a polymer is a chain made up of repeating polymers.

B is incorrect because the group of four atoms described is part of an organic acid.

C is correct because the four atoms bonded in this way form a carboxyl group.

D is incorrect because an aromatic compound is a compound containing carbon rings.

4. D

A is incorrect because carbon generally bonds with halogens. Silver is not a halogen.

B is incorrect because helium is not commonly found bonded to carbon in organic molecules.

C is incorrect because sodium is a reactive metal, but carbon bonds with a different group of elements.

D is correct because carbon often bonds to itself and certain other elements that include hydrogen and the halogens.

5. D

A is incorrect because many elements are solids at room temperature yet cannot form the diverse compounds that carbon can.

B is incorrect because several other elements are also nonmetals, but they are not as unique as carbon is.

C is incorrect because many elements have different forms according to the number of neutrons in their atoms.

D is correct because each carbon atom has four valence electrons and can therefore bond with up to four other atoms.

Lesson 3 Quiz
1. B   4. D
2. C   5. B
3. D

1. B

A is incorrect because producing and sustaining the conditions necessary for fusion are challenging and expensive.

B is correct because fusion reactions do not produce radioactive waste.

C is incorrect because the energy produced by fusion does need to be contained, and containing nuclear fusion is currently a challenging task.

D is incorrect because this would be a disadvantage of using nuclear fusion over nuclear fission.

2. C

A is incorrect because chemical reactions can also release energy.

B is incorrect because nuclear reactions and chemical reactions can release energy.

C is correct because changes to the nucleus of an atom can alter the number of protons and change the identity of the atom.

D is incorrect because nuclear reactions take place in the sun and other stars.

3. **D**

A is incorrect because nuclear power plants are powered by fission reactions. Beta decay is not an example of a fission reaction.

B is incorrect because nuclear power plants are powered by fission reactions. Gamma decay is not an example of a fission reaction.

C is incorrect because nuclear power plants are powered by fission reactions. Nuclear fusion is being investigated, but it is not currently in use in nuclear power plants.

D is correct because nuclear power plants are powered by controlled fission chain reactions.

4. **D**

A is incorrect because energy is given off during alpha decay.

B is incorrect because alpha particles are emitted during alpha decay. The alpha particle is shown at the top right.

C is incorrect because alpha decay releases two smaller nuclei, one of which is an alpha particle.

D is correct because gamma rays are given off during gamma decay, not alpha decay.

5. **B**

A is incorrect because atoms of the same element must have the same atomic number.

B is correct because isotopes have different mass numbers due to a difference in the number of neutrons.

C is incorrect because ion is the term that describes atoms that gain or lose electrons to acquire a positive or negative charge.

D is incorrect because atoms of the same element, including different isotopes, have the same number of electrons.

Lesson 1 Alternative Assessment

Vocabulary: Students use the words *chemical reaction, chemical formula, chemical equation, reactant, product, law of conservation of mass, endothermic reaction, exothermic reaction, law of conservation of energy* to write sentences that are appropriate and accurate.

Examples: Students take several pictures that show chemical reactions. Then they write or draw descriptions that explain what the chemical reaction was and whether it was an endothermic or exothermic reaction.

Analysis: Students explain how engines that we use for transportation transform energy from one form to another. Explanations should identify how energy is changed, but not lost or gained in these processes.

Observations: Students use a paper cup, a pad of steel wool, and vinegar to demonstrate a chemical reaction and then explain what caused the reaction.

Calculations: Students list some chemical reactions. Then they write one of the reactions in the form of a balanced chemical equation.

Lesson 2 Alternative Assessment

In Your Words: An organic compound is a compound that is based on the element carbon.

Why Carbon? Carbon's ability to form long chains and bond with other elements makes it an important compound in the chemistry of living organisms.

Identifying Relationships: Five of the six elements that make up most organic compounds—carbon, nitrogen, oxygen, phosphorus, and sulfur—are all next to each other on the periodic table and are nonmetals.

Paper Chains: Make sure students create three different types of carbon molecules. Check to see that the molecule is labeled correctly.

Gum Drop Molecules: Check to make sure student models accurately represents a straight, double, and triple bond.

Diagramming Bonds: Remind students that each carbon atom can form a total of four bonds.

Carbon dioxide: O=C=O; ethane: the carbon atoms single-bonded to each other, and each carbon atom is single-bonded to three hydrogen atoms.

What and Where: The six most common elements in living things are carbon, hydrogen, oxygen, nitrogen, sulfur, and phosphorus. The human body contains water and water is made up of hydrogen and oxygen. Nitrogen and sulfur are important parts of proteins, and phosphorus is found in every human cell.

Writing from Research: Answers should include pertinent information on the compound's discovery, relationships between the compound's structure and function, and the compound's economic value and importance.

Polymer Posters: Posters should include the name of a polymer and explain how this polymer improves a particular product.

Nutrition Report: Reports should include how an organic compound influences human health.

Lesson 3 Alternative Assessment

Illustrating Isotopes: Student diagrams clearly show that the nucleus of carbon-12 contains 6 protons and 6 neutrons and the nucleus of carbon-14 contains 6 protons and 8 neutrons. Check labels for accuracy.

Making Models: Student models correctly show the differences between alpha decay, beta decay, and gamma decay.

Decay Poster: Student posters correctly compare alpha, beta, and gamma particles. Captions and labels clearly show the differences between each type of radiation.

Model It! Models correctly show the number of protons and neutrons that make up a U-235 nucleus and indicate that during fission the nucleus splits to form a krypton-91 nucleus and a barium-142 nucleus. Models also show that three neutrons are released.

Concept Mapping: Concept maps correctly explain nuclear fission and nuclear fusion.

Radiation in Use: Posters or visual displays clearly show at least one use of radiation for each category. Possible illustrations will include: home: smoke detector; industry: find leaks in pipes or deter-mine thickness of metals; medicine: use as tracers, for X-rays, to diagnose or treat illness; energy source: nuclear power plant.

Performance-Based Assessment

See Unit 4, Lesson 1
2. Nothing happens.
3. The solution becomes cloudy, and a white precipitate forms.
4. Nothing happens.
5. The solid dissolves, and gas bubbles form.
6. The solution turns pink.
7.

		Liquid A	Liquid B
Chemical Reactions	Liquid B	no reaction	
	Liquid C	white precipitate	
	Liquid D	no reaction	solution turns pink
	Solid E	gas bubbles	

8. A reaction may or may not have taken place, but there is no visible evidence of a reaction.
9. A reaction did take place because a precipitate formed.
10. Some type of gas is produced during this reaction.
11. The properties of the new substance are different from the properties of the original substances.

Unit Review
Vocabulary
1. T See Unit 4, Lesson 1
2. F See Unit 4, Lesson 3
3. F See Unit 4, Lesson 3
4. F See Unit 4, Lesson 1
5. F See Unit 4, Lesson 2

Key Concepts
6. B 9. B 12. A
7. D 10. B 13. B
8. D 11. D

6. B See Unit 4, Lesson 3
A is incorrect because a catalyst is used to speed a reaction and is not a type of atom.
B is correct because an isotope of an atom has a different number of neutrons and thus a different mass number.
C is incorrect because a reactant is something that reacts in a chemical reaction.
D is incorrect because a product is something that is produced in a chemical reaction.

7. D See Unit 4, Lesson 1
A is incorrect because a catalyst is used to speed a reaction.

B is incorrect because lowering the reactant concentration may slow a reaction.

C is incorrect because higher temperatures generally increase the rate of reaction.

D is correct because increasing surface area exposes more of the reactant and increases the rate of reaction.

8. **D See Unit 4, Lesson 3**

A is incorrect because nuclear fusion does not produce radioactive waste.

B is incorrect because hydrogen fuel can be obtained from Earth's oceans.

C is incorrect because instruments have been developed to work with atoms and subatomic particles.

D is correct because extreme pressure and temperature, such as the conditions on the sun, are needed to produce nuclear fusion.

9. **B See Unit 4, Lesson 1**

A is incorrect because there is one molecule, but two atoms of carbon.

B is correct because the subscript to the right of a chemical symbol indicates how many atoms of a given element are present.

C is incorrect because there are three atoms of hydrogen, but two of carbon.

D is incorrect because there are 7 atoms total in one molecule of acetate, but only two of them are carbon atoms.

10. **B See Unit 4, Lesson 2**

A is incorrect because a hydrocarbon contains only carbon and hydrogen.

B is correct because carbohydrates are neutral compounds made of carbon, hydrogen, and oxygen.

C is incorrect because acids do not necessarily contain carbon, hydrogen, or oxygen, and they are not always molecules.

D is incorrect because a polymer is any long chain carbon molecule with a repeating structure.

11. **D See Unit 4, Lesson 1**

A is incorrect because Fe_3O_4 is only one of the two products produced.

B is incorrect because Fe is one of the reactants.

C is incorrect because Fe and H_2O are the reactants.

D is correct because products are the substances that are formed from a chemical reaction. They are written to the right of the yields sign.

12. **A See Unit 4, Lesson 1**

A is correct because an odor is a sign that a chemical reaction has taken place, and burning a sugar cube produces a new substance.

B is incorrect because melting an ice cube is a physical change.

C is incorrect because crushing a can is a physical change.

D is incorrect because breaking a bottle is a physical change.

13. **B See Unit 4, Lesson 1**

A is incorrect because photosynthesis is an endothermic reaction that takes place in the cells of photosynthetic organisms.

B is correct because burning wood releases light and energy in the form of heat.

C is incorrect because melting ice cubes is a physical change that requires the input of energy.

D is incorrect because boiling water is a physical change that requires the input of energy.

Critical Thinking

14. **See Unit 4, Lesson 3**

- identifies fission
- expresses that the nucleus of an atom splits into two smaller nuclei, releasing smaller particles
- recognizes that the 3 neutrons released from the first reaction could each split additional atoms of uranium-235 in a cascade effect

15. **See Unit 4, Lesson 3**

- names alpha particles, beta particles, and gamma particles
- mentions use of radioactive decay in medical imaging, medical treatment, or dating artifacts
- clearly expresses risks and benefits of using radioactive decay using supportive evidence and describes opinion as to whether the benefits outweigh the risks (e.g. *The use of radioactive decay in*

imaging helps save more lives through diagnoses than it endangers; etc.)

Connect Essential Questions

16. See Unit 4, Lesson 1 and Lesson 2

- balances the equation by placing the coefficient 2 in front of H_2O
- expresses that matter cannot be created or destroyed and that atoms must be accounted for on each side of a balanced equation
- identifies CH_4 as a hydrocarbon
- draws a full structural drawing of CH_4 showing four hydrogen (H) atoms bonded to the carbon atom (C).

Unit Test A

Key Concepts

1. A   5. C   9. A
2. C   6. D   10. B
3. D   7. B   11. A
4. B   8. A   12. A

1. A

A is correct because potatoes contain complex carbohydrates.

B is incorrect because many simple sugars come together to form complex carbohydrates.

C is incorrect because corn syrup contains simple sugars.

D is incorrect because simple sugars must come together to form complex carbohydrates.

2. C

A is incorrect because carbon is an atom in two of the substances, but it is not a product by itself.

B is incorrect because oxygen is a reactant in the equation.

C is correct because water is produced in this reaction.

D is incorrect because ethane enters into the reaction as a reactant.

3. D

A is incorrect because energy can also be released by reactions that are not nuclear reactions.

B is incorrect because electrons are shared in some types of chemical bonds but not in nuclear reactions.

C is incorrect because electrons are transferred in the formation of some chemical bonds but not during nuclear reactions.

D is correct because nuclear reactions affect the nucleus of an atom.

4. B

A is incorrect because mass is conserved in all chemical reactions.

B is correct because burning is an example of an exothermic reaction. Thermal energy and light are released.

C is incorrect because energy is released, not absorbed, during burning.

D is incorrect because there must be at least one product formed.

5. C

A is incorrect because a catalyst would increase the rate of a reaction.

B is incorrect because lowering the temperature would likely decrease the rate of a chemical reaction.

C is correct because increasing surface area would increase the number of collisions between reactant particles.

D is incorrect because decreasing the concentration would decrease the rate of a chemical reaction.

6. D

A is incorrect because there are six atoms of carbon alone.

B is incorrect because there are 12 atoms of hydrogen alone.

C is incorrect because there are 18 atoms of carbon and hydrogen combined, but this number does not include oxygen.

D is correct because there are a total of 24 atoms in this molecule.

7. B

A is incorrect because MgO is a product in an equation, not a reactant.

B is correct because Na_3PO_4 is a product in the first equation.

C is incorrect because the subscript indicates three atoms of hydrogen, not oxygen.

D is incorrect because catalysts do not change to form other

Answer Key

substances in a chemical reaction.

8. A

A is correct because carbon molecules may form long straight chains, branched chains, or rings.

B is incorrect because carbon molecules may form branched chains.

C is incorrect because carbon molecules can take other forms.

D is incorrect because many carbon molecules contain many carbon atoms; some molecules contain hundreds.

9. C

A is incorrect because a neutron becomes several other subatomic particles, including a proton, so the mass number does not change.

B is incorrect because a neutron becomes several other subatomic particles, including a proton, so the mass number does not change.

C is correct because the atomic number increases as the number of protons increase; the result of this decay would be a nitrogen atom.

D is incorrect because the atomic number increases as the number of protons increase; the result of this decay would be a nitrogen atom.

10. A

A is correct because a small amount of mass can produce a large amount of energy.

B is incorrect because the sun does not affect the energy released during fission reactions.

C is incorrect because water is not the energy source of a nuclear fission reaction.

D is incorrect because gamma rays are not produced from electrons.

11. B

A is incorrect because the diagram shows more than one carbon atom.

B is correct because the structure shows 2 carbon atoms and 6 hydrogen atoms.

C is incorrect because the structural formula shows the number of each type of atom. The structural formula should be the same as the chemical formula.

D is incorrect because the number of hydrogen atoms is different from the number of carbon atoms.

12. A

A is correct because unstable atoms decay until they become stable.

B is incorrect because isotopes are different forms of the same element.

C is incorrect because nuclei break apart during radioactive decay to form smaller nuclei.

D is incorrect because atoms do not gain electrons during radioactive decay.

Critical Thinking
13.
- long-chain carbon molecules composed of repeated structural units called monomers
- monomers
- examples of polymers (e.g., *Examples of polymers include nylon, Kevlar, and keratin*; etc.)

Extended Response
14.
- 4 and 2+
- description of a nuclear fusion reaction (e.g., *The nuclei of two smaller atoms come together to form a larger nucleus*; etc.)
- explanation of energy production in fusion reaction (e.g., *During a nuclear fusion reaction, a small amount of mass is converted to a large amount of energy*; etc.)

Unit Test B
Key Concepts
1. B   5. C   9. D
2. C   6. D   10. A
3. B   7. C   11. D
4. A   8. B   12. B

1. B

A is incorrect because complex carbohydrates and simple sugars are made up of the same kinds of atoms.

B is correct because many simple sugars come together to form complex carbohydrates.

C is incorrect because the ratio of carbon to hydrogen to

oxygen is the same in both kinds of compounds.

D is incorrect because carbohydrates and simple sugars are made up of carbon, hydrogen, and oxygen.

2. **C**

A is incorrect because the reactants are correctly to the left of the arrow in the equation, and the products are correctly to the right of the arrow.

B is incorrect because the formulas of all reactants (ethane and oxygen) and all products (carbon dioxide and water) are correct.

C is correct because the numbers of each type of atom is not the same on both sides of the equation.

D is incorrect because both sides of the equation show the presence of C, H, and O atoms.

3. **B**

A is incorrect because energy can be emitted without changing the structure of the nucleus.

B is correct because protons determine the identity of the atom.

C is incorrect because atoms of the same element can have different numbers of neutrons.

D is incorrect because chemical reactions involve valence electrons, which do not determine the identity of the atom. Nuclear reactions involve changes to the nucleus of an atom.

4. **A**

A is correct because the reaction that happens when a candle burns releases energy.

B is incorrect because the reaction of burning does not absorb energy; it releases energy.

C is incorrect because energy is not destroyed during any chemical changes.

D is correct because energy is required in order for the chemical reaction of burning to proceed.

5. **C**

A is incorrect because the temperature changes the amount of any component present during a reaction, not only the catalyst.

B is incorrect because temperature does not affect the number of reactant particles.

C is correct because lowering the temperature decreases the average speed of the particles, so they collide less often.

D is incorrect because temperature does not alter the sizes of the reactant particles.

6. **D**

A is incorrect because the coefficient and subscript must be multiplied to find the number of carbon atoms. 2 is the subscript only.

B is incorrect because the coefficient and subscript must be multiplied to find the number of carbon atoms. 3 is the coefficient only.

C is incorrect because the coefficient and subscript must be multiplied to find he number of carbon atoms. 5 is the result of adding the coefficient and subscript.

D is correct because the coefficient and subscript must be multiplied to find the number of carbon atoms in the formula.

7. **C**

A is incorrect because mass is conserved during any chemical reaction, and the reactants and products are balanced in this equation.

B is incorrect because bonds were broken between the oxygen atoms in the oxygen molecule to form the new compound of MgO.

C is correct because the chemical properties of products are different from the chemical properties of the reactants.

D is incorrect because the chemical properties of reactants are different from the chemical properties of the products.

8. **B**

A is incorrect because the number of protons in a carbon atom's nucleus is not directly related to its ability to form so many different types of compounds.

B is correct because carbon atoms can make four bonds with up to four other atoms.

C is incorrect because other elements are also capable of radioactive decay.

D is incorrect because other elements also exist in nature in solid, liquid, and gaseous states.

9. **D**

A is incorrect because gamma rays have no mass.

B is incorrect because no particles are released in this form of decay.

C is incorrect because gamma rays have no charge or mass.

D is correct because gamma rays are a form of energy.

10. **A**

A is correct because a small amount of mass can produce a large amount of energy when it splits through the process of nuclear fission.

B is incorrect because the nucleus of an atom does not absorb electrons from other atoms.

C is incorrect because nuclei do not form bonds with each other.

D is incorrect because nuclear fusion happens when the nuclei of smaller atoms combine to form a new, more massive nucleus.

11. **D**

A is incorrect because the structure shows 6 carbon atoms and 14 hydrogen atoms.

B is incorrect because the diagram shows more than 12 atoms.

C is incorrect because the structural formula shows the number of each type of atom. The structural formula should contain the same numbers as the chemical formula.

D is correct because the structural formula shows 6 carbon atoms and 14 hydrogen atoms.

12. **B**

A is incorrect because some elements are stable and do not decay.

B is correct because some elements are unstable because of the forces that exist within the nucleus.

C is incorrect because naturally occurring elements can also decay.

D is incorrect because having more neutrons than protons does not guarantee decay.

Critical Thinking

13.
- an organic compound that has acid properties
- a combination of four atoms consisting of carbon, oxygen, and an -OH group
- description of structural diagram of carboxyl group (e.g., *I could draw a capital C connected by a double line on one side to a capital O, a single line on another side to a capital OH, and a single line on a third side to another molecule*; etc.)
- examples of organic acids (e.g., *Examples of organic acids include fatty acids,*

amino acids, and nucleic acids; etc.)

Extended Response

14.
- 4 and 2+
- description of a nuclear fusion reaction (e.g., *The nuclei of two smaller atoms come together to form a larger nucleus*; etc.)
- explanation of energy production in fusion reaction (e.g., *During a nuclear fusion reaction, a small amount of mass is converted to a large amount of energy*; etc.)
- identification of fusion examples that happen in nature (e.g., *The fusion of hydrogen in the sun releases the energy upon which Earth depends*; etc.)

Unit 5 Solutions, Acids, and Bases

Unit Pretest

1. A   5. D   9. B
2. A   6. C   10. B
3. A   7. C
4. A   8. C

1. **A**

A is correct because a very high pH means that a compound is highly basic.

B is incorrect because highly acidic compounds have a very low pH.

C is incorrect because slightly basic compounds have a pH slightly above 7.

D is incorrect because slightly acidic compounds have a pH slightly below 7.

Answer Key

2. A

A is correct because an acid and a base combine to form water and a salt during a neutralization reaction.

B is incorrect because a base is a reactant in a neutralization reaction.

C is incorrect because an acid enters into a neutralization reaction.

D is incorrect because a metal replaces a hydrogen in an acid. The metal is not a product.

3. A

A is correct because the concentration is found by dividing the mass of the solute by the volume of the solvent, and then multiplying by 100%.

B is incorrect because the concentration is not necessarily the same magnitude as the mass of the solute.

C is incorrect because the concentration is the same as 40 g of sugar per 100 mL of solution. The concentration is not found by subtracting the mass of the solute from the volume of the solution.

D is incorrect because the concentration compares the amount of solute (mass) to the amount of solvent (volume). The concentration is not found by adding the mass of the solute and the volume of the solution.

4. A

A is correct because solutions do not scatter light, whereas colloids and suspension do. The particles in suspensions are larger than in colloids and the particles therefore become visible.

B is incorrect because the particles in a solution are too small to scatter light.

C is incorrect because the colloid and the suspension would scatter light and the first container in the image does not show scattered light, indicative of a solution.

D is incorrect because a colloid has smaller particles than a suspension does.

5. D

A is incorrect because salts are not used to identify acids and bases.

B is incorrect because acid-base indicators typically use color change for identification. Tasting would not be a safe method.

C is incorrect because acids and bases are corrosive. However, this property is not a factor in acid-base indicators.

D is correct because acid-base indicators turn specific colors in the presence of either an acid or a base.

6. C

A is incorrect because acid rain alters the composition of rain, but it does not lead to the production of additional rain.

B is incorrect because acid rain does not alter the temperature of normal rain.

C is correct because the acid rain has a lower pH than normal rain. This can harm fish and other aquatic life.

D is incorrect because acid rain changes the chemical composition of rain, but it does not destroy rain.

7. C

A is incorrect because antacids neutralize stomach acids, so antacids are bases.

B is incorrect because bases are used as soaps and detergents.

C is correct because salts, such as calcium carbonate, are used to melt the ice on sidewalks and roadways.

D is incorrect because bases, such as ammonia, are commonly used in fertilizers.

8. C

A is incorrect because a salt would not break up into H+ ions as shown.

B is incorrect because a base would form hydroxide ions.

C is correct because an acid would increase the number of hydrogen ions in solution.

D is incorrect because a solution with hydrogen ions, as shown, is formed from a specific type of compound.

9. B

A is incorrect because adding more water will increase the volume, but not cause the tablet to break apart more quickly.

B is correct because shaking will lead to more collisions

between solvent and solute particles.

C is incorrect because larger tablets will decrease the surface area of tablet particles, thereby slowing the rate of solution.

D is incorrect because lowering the temperature will make the tablet dissolve more slowly.

10. B

A is incorrect because pH is not a measure of volume.

B is correct because the pH value of a solution is a measure of the solution's hydrogen ion concentration.

C is incorrect because the pH value does not depend on the size of the particles.

D is incorrect because the average speed of the particles is related to temperature rather than pH.

Lesson 1 Quiz
1. B   4. A
2. D   5. A
3. B

1. B

A is incorrect because the gas solute escapes the liquid solvent. This change is not a function of the speed at which a solute dissolves.

B is correct because solubility is related to pressure. Increasing pressure increases the solubility of gases in liquids, whereas decreasing pressure by opening a beverage decreases the solubility of gases in liquids.

C is incorrect because gas bubbles are not released

because the liquid suddenly becomes saturated.

D is incorrect because the gas is the solute and the liquid beverage is the solvent. Opening the beverage changes the pressure and not the state of the liquid.

2. D

A is incorrect because dilute solutions can have solutes in any state of matter.

B is incorrect because both kinds of solutions can be at the same temperature.

C is incorrect because the volume of a dilute solution can be greater than that of a concentrated solution.

D is correct because a dilute solution has a low concentration of solute, whereas a concentrated solution has a high concentration of solute.

3. B

A is incorrect because salt would be considered the solute in this solution.

B is correct because the salt (solute) dissolves in the water (solvent).

C is incorrect because the salt and water together form a mixture called a solution.

D is incorrect because a suspension is a different type of mixture in which the particles are larger and remain suspended in the liquid.

4. A

A is correct because the sugar and water form a solution.

B is incorrect because the sugar (solute) mixes thoroughly into the water (solvent).

C is incorrect because the sugar and water form a solution. The particles in solutions are so small they will not scatter light as a suspension or a colloid would.

D is incorrect because the water can be evaporated to leave the sugar behind. A chemical change has not occurred.

5. A

A is correct because cooling the solvent will decrease the solubility.

B is incorrect because crushing the sugar will increase the surface area exposed to the solvent. This will increase the solubility.

C is incorrect because increasing the amount of sugar will not affect how quickly it dissolves.

D is incorrect because stirring will cause the solute to dissolve more quickly in the solvent.

Lesson 2 Quiz
1. B   4. A
2. C   5. B
3. B

1. B

A is incorrect because a salt does not have a bitter taste. It has a characteristic salty taste, and it is not slippery.

B is correct because bases are generally bitter, slippery, and conduct an electric current.

C is incorrect because acids taste sour rather than bitter, and are not slippery.

D is incorrect because distilled water has no characteristic taste and is not slippery.

2. **C**

A is incorrect because the reaction between a hydrogen ion and a hydroxide ion forms water.

B is incorrect because a salt forms from a chemical change rather than a physical change.

C is correct because the negative ion of an acid and the positive ion of a base combine to form a salt.

D is incorrect because although water is broken into hydroxide and hydrogen ions, in a neutralization reaction, these ions form water in addition to a salt being formed.

3. **B**

A is incorrect because a base breaks apart to release hydroxide ions into water.

B is correct because an acid donates hydrogen ions, which combine with water molecules to form hydrogen ions.

C is incorrect because a salt is formed when an acid reacts with a base, not with water.

D is incorrect because water naturally breaks apart into hydrogen ions and hydroxide ions rather than individual atoms.

4. **A**

A is correct because a neutralization reaction is a reaction between an acid and a base to form a salt and water.

B is incorrect because oxygen alone is not a product of a neutralization reaction.

C is incorrect because the hydrogen ion is supplied by the acid, but it combines with the hydroxide ion from the base to form water.

D is incorrect because the hydroxide ion is supplied by the base, but it combines with the hydrogen ion from the acid to form water.

5. **B**

A is incorrect because a different substance, called hemoglobin, is responsible for this task.

B is correct because hydrochloric acid breaks down foods in the stomach during digestion.

C is incorrect because the kidneys filter wastes from the blood without the use of hydrochloric acid.

D is incorrect because certain types of blood cells kill pathogens as part of the immune system. Hydrochloric acid is used in the digestive system.

Lesson 3 Quiz
1. A   4. C
2. A   5. B
3. B

1. **A**

A is correct because absorbing carbon dioxide gas makes ocean water more acidic, thereby reducing its pH.

B is incorrect because carbon dioxide gas dissolves in the water. It changes the solution in a different way.

C is incorrect because no water is destroyed when the gas is absorbed.

D is incorrect because carbon dioxide does not produce additional salt in the water.

2. **A**

A is correct because the pH value of a solution is a measure of its hydrogen ion concentration.

B is incorrect because a different value is used to measure pH.

C is incorrect because the reactivity of an element describes how likely it is to take part in a chemical reaction. The pH value describes something about a solution.

D is incorrect because the ratio of solute to solvent can be used to measure concentration.

3. **B**

A is incorrect because solutions with lower pH values are acids.

B is correct because bases have pH values from 7 to 14, with the stronger bases having higher values.

C is incorrect because pH is used to describe alkalinity (basicity)

D is incorrect because pH and alkalinity (basicity), although

Answer Key

related, are different concepts.

4. **C**

A is incorrect. Oil spills do harm the environment, but not by producing acid rain.

B is incorrect because plants release oxygen into the air, which living things need.

C is correct because the burning of fossil fuels produces gases that form acid rain in the atmosphere.

D is incorrect because warm water released into waterways can harm living things, but not by producing acid rain.

5. **B**

A is incorrect because lye is extremely basic. This location is too far along the scale.

B is correct because a pH of 7 represents a neutral substance. Human blood would be just above this pH value because it is a weak base.

C is incorrect because lemon juice is a strong acid. Human blood is a base.

D is incorrect because human blood should be located closer to a neutral pH, or 7 on the scale.

Lesson 1 Alternative Assessment

Terms: Paragraphs clearly and accurately define *solution, solute, solvent, concentration,* and *solubility* and demonstrate the relationship between them.

Paragraphs are cohesive and flow.

Examples: Students accurately identify the following types of solutions: liquid in liquid, liquid in solid, liquid in gas, solid in solid, solid in gas, and gas in gas solutions. Students also provide correct examples for each type.

Illustrations: Diagrams or illustrations show the difference between solutions, colloids, and suspensions clearly and accurately.

Analysis: Students correctly identify the solute (copper sulfate) and solvent (water), and identify factors that affect solubility.

Models: Students correctly model how a solution is made, including how temperature and pressure affect solubility.

Lesson 2 Alternative Assessment

Staying Neutral: Students should draw a diagram or illustration that shows a neutralization reaction.

Pass the Salt! Students should write a paragraph in which they identify common uses for and examples of salts.

Base Hit: Students should write a song, poem, or paragraph that tells the properties of bases.

Base Camp: Students should write and perform a skit that demonstrates how bases break up in water.

Acid Looks: Students should prepare a multimedia presentation shows how acids break up in water.

Acids By the Book: Students should write an illustrated booklet that identifies the physical and chemical properties of acids.

Lesson 3 Alternative Assessment

The pH Scale: The pH scale should include numbers from 1 to 14, with items placed as follows: pure water, 7; normal rain water, 5 to 6; lemon juice, 2; baking soda, 9; soap, 10; and acid rain, 4.5 to 5.

Difference in pH: A change from pH 9 to pH 5 is 4 units down on the pH scale. The concentration of hydronium ions would be $10 \times 10 \times 10 \times 10$, or 10,000 times greater at pH 5 than at pH 9.

Soil pH: Columns explain that hydrangea flowers will be blue if the soil is acidic and pink if it is neu-tral or basic. Columns also describe specific techniques for amending soil to adjust pH, such as add-ing lime to decrease acidity.

Stomach pH: Pamphlets explain causes of indigestion and how antacid tablets can be used to in-crease the pH of the stomach. Pamphlets are written in clear, easily understood language, appropriate for a non-technical audience.

Measuring pH: Posters show how to use universal pH paper, an acid-base indicator, and an electronic pH meter. Students should identify the electronic meter as the most accurate.

In Our Bodies: Students should identify three causes of acidosis or alkalosis, how health is adversely affected by each condition, and how the body responds to restore healthy pH.

Acid Rain: Diagrams correctly identify how the burning of fossil fuels causes acid rain and three effects of acid rain. Students also explain two ways people can use less fossil fuels.

Performance-Based Assessment

See Unit 5, Lesson 1

1. Sample answer: The solubility of a solute increases when the temperature of the solvent increases.

2. Sample answer: I predict that more salt will dissolve in the hot water than in the cold water.

7. See table below. These are sample data. Student data may vary.

Amount of Dissolved Salt

Cold water (g)	Hot water (g)
6	8

8. The hot water dissolved more salt.

9. Student data should show that more salt dissolves in hot water than in cold water. Be sure students use their real data to answer the question.

10. Answers may vary. Accept all reasonable answers. Student data should show that more salt dissolves in hot water than in cold water. Be sure students use their real data to answer the question.

11. Salt is the solute because it does the dissolving. Water is the solvent because the solute dissolves into it. The salt water is the solution because it contains both the solvent and the dissolved solute.

Unit Review

Vocabulary
1. F See Unit 5, Lesson 2
2. T See Unit 5, Lesson 3
3. T See Unit 5, Lesson 1
4. T See Unit 5, Lesson 1
5. F See Unit 5, Lesson 2

Key Concepts
6. D   9. A   12. A
7. C   10. B   13. A
8. B   11. C

6. **D See Unit 5, Lesson 3**

   A is incorrect because ammonia is basic and will have a pH value greater than 7.

   B is incorrect because lemon juice is acidic and will have a pH value less than 7.

   C is incorrect because ammonia is basic and lemon juice is acidic.

   D is correct because ammonia is basic and lemon juice is acidic.

7. **C See Unit 5, Lesson 3**

   A is incorrect because antacids help neutralize stomach acids.

   B is incorrect because antacids help make stomach acids more basic.

   C is correct because antacids help neutralize stomach acids to help with conditions such as acid re-flux.

   D is incorrect because neutralizing stomach acids won't help to release more nutrients.

8. **B See Unit 5, Lesson 3**

   A is incorrect because there is only a 10 fold decrease in hydroniumion concentration between each increasing increment on the pH scale.

   B is correct because there is a 100 fold decrease in hydronium concentration between the two increments on the pH scale (between 2-4).

   C is incorrect because a solution with a pH between 8 and 10 on the pH scale is basic.

   D is incorrect because 9-12 represents a 1,000 fold increase in hydronium concentration, and describes a solution that is basic.

9. **A See Unit 5, Lesson 1**

   A is correct because the particles in a solution do not scatter light.

   B is incorrect because light is scattered in glass B and solutions do not scatter light.

   C is incorrect because glass B is a mixture, such as a suspension, that scatters light.

   D is incorrect because one property of solutions is that the particles do not scatter light.

Answer Key

10. B See Unit 5, Lesson 1

A is incorrect because the surface area of the solute, not the solvent, is increased.

B is correct because it allows more particles of solute to come into contact with the solvent.

C is incorrect because the temperature is not appreciably increased through crushing or stirring.

D is incorrect because stirring or crushing a solute does not necessarily result in an increased pressure of the solvent.

11. C See Unit 5, Lesson 3

A is incorrect because a saturated solution has the maximum amount of solute that can be dissolved in the solvent.

B is incorrect because a concentrated solution has a high concentration of solute.

C is correct because a dilute solution has a low concentration of solute.

D is incorrect because a suspension is a type of mixture but is not a solution.

12. A See Unit 5, Lesson 1

A is correct because the solute is the substance that dissolves in the solvent.

B is incorrect because water is the solvent in a sugar solution.

C is incorrect because the beaker is the container for the sugar solution.

D is incorrect because a solution is the result of dissolving a solute in a solvent.

13. A See Unit 5, Lesson 3

A is correct because both pH paper and indicators change color to indicate pH value of a solution.

B is incorrect because the shape of the indicator does not change when pH is measured.

C is incorrect because, although a number scale is used, it is the color of the indicator that changes when pH is measured by these methods.

D is incorrect because there are no letters associated with pH measurement.

Critical Thinking
14. See Unit 5, Lesson 2

- identifies an acid (e.g., *sulfuric, nitric, hydrochloric, hydrofluoric*; etc.)
- names a use of that acid (e.g., *industrial processes, cleaning, break down food*; etc.)
- identifies a base (e.g., *ammonia, magnesium hydroxide*; etc.)
- names a use of that base (e.g., *soap, detergents, drain cleaner, fertilizer, antacids*; etc.)

15. See Unit 5, Lesson 1

- identifies apple juice as a solution
- expresses that a solution is homogeneous, or that the substances are uniformly dispersed; the properties of a solution are the same in every part of a solution
- recognizes that filtering would not separate the particles of the apple juice

Connect Essential Questions
16. See Unit 5, Lesson 2 and Lesson 3

- identifies that burning fossil fuels produce emissions that combine with water in the atmosphere and fall as acid rain
- expresses that acids are corrosive, meaning that they can react with and destroy materials
- identifies that acid rain lowers the pH of soil and water environment
- recognizes that acid rain can harm organisms (e.g. *Acid rain can kill plants and trees and harm fish and other wildlife*; etc.)
- reasons that reducing or eliminating pollution, especially the burning of fossil fuels, would reduce acid rain

Unit Test A
Key Concepts
1. D   5. A   9. B
2. C   6. C   10. A
3. B   7. A   11. C
4. D   8. B   12. D

1. D

A is incorrect because vinegar is a strong acid.

B is incorrect because pure rain is slightly acidic.

C is incorrect because household lye is a strong base.

D is correct because distilled water has a neutral pH of 7. It is neither an acid nor a base.

2. C

A is incorrect because a salt will not reduce the pH of a substance.

B is incorrect because a base would raise the pH of ocean water.

C is correct because carbon dioxide makes ocean water more acidic.

D is incorrect because a neutral substance would not affect the pH of ocean water.

3. B

A is incorrect because the acetic acid is the solute.

B is correct because water does the dissolving.

C is incorrect because solubility describes the ability of one substance to dissolve in another.

D is incorrect because concentration relates the amount of acetic acid to water in the solution.

4. D

A is incorrect because the line for KCl slants upward, which means that its solubility increases as temperature increases.

B is incorrect because the line for $NaNO_3$ has a positive slope, which means that more solute can dissolve at higher temperatures.

C is incorrect because the line for $KClO_3$ curves upward, which means that the amount of solute that can dissolve increases as temperature increases.

D is correct because the line for $Ce_2(SO_4)_3$ curves downward and then levels out. Less solute can dissolve at 40°C, for example, than at 10°C.

5. A

A is correct because nitric acid is produced when fossil fuels are burned.

B is incorrect because ammonia is a base and will not lower the pH of rain.

C is incorrect because citric acid is what gives oranges, lemons, and other citrus fruits their sour taste.

D is incorrect because hydrochloric acid is not released into the air as a gas. One place hydrochloric acid is found is in the human stomach.

6. C

A is incorrect because a neutralization reaction produces a salt.

B is incorrect because a neutralization reaction does not involve sugar.

C is correct because the base would react with the stomach acid in a neutralization reaction.

D is incorrect because an acid is not neutralized by another acid.

7. A

A is correct because chalk, which is calcium carbonate, is a salt.

B is incorrect because bleach, soaps, and other cleaning agents contain bases.

C is incorrect because ammonia is a base.

D is incorrect because milk of magnesia is a base commonly used to neutralize acids.

8. B

A is incorrect because the same type of mixture can have different numbers of particles.

B is correct because the mixtures are distinguished by the size of the particles. The particles are smallest in solutions and largest in suspensions.

C is incorrect because many mixtures have no color at all.

D is incorrect because temperature does not separate the three types of mixtures.

9. B

A is incorrect because the scientist would add solvent to dilute the solution.

B is correct because bromthymol blue is an acid-base indicator.

C is incorrect because concentration relates the amount of solute to solvent.

D is incorrect because bromthymol blue does not separate the parts of a solution.

10. A

A is correct because a base breaks apart to release hydroxide ions into water.

B is incorrect because an acid donates hydrogen ions to water, which forms hydrogen ions.

C is incorrect because a salt is formed when an acid reacts with a base, not with water.

D is incorrect because water naturally breaks apart into hydrogen ions and hydroxide ions rather than into individual atoms.

11. C

A is incorrect because heating may help the salt dissolve faster, but it will not make the solution more concentrated.

B is incorrect because shaking may help the salt dissolve faster, but it will not make the solution more concentrated.

C is correct because a concentrated solution has more solute per volume of solvent than a dilute solution has.

D is incorrect because adding water will dilute the solution further.

12. D

A is incorrect because although some foods contain acids and bases, they are not nutrients that living things need to stay alive. They are used by living things to obtain some nutrients that the organism needs.

B is incorrect because acids and bases do not form coatings that protect other materials.

C is incorrect because many acids and bases are found naturally.

D is correct because acids and bases can react with body tissues, clothing, and other materials.

Critical Thinking
13.
- definition of solubility (e.g., *Solubility is the ability of one substance to dissolve in another at a given temperature and pressure*; etc.)
- explanation relating solubility to saturation (e.g., *If the solute has a low solubility, the solution can become saturated with only a small amount of solute dissolved*; etc.)

Extended Response
14.
- solution A
- explanation of what makes a solution acidic (e.g., *A solution with a higher concentration of hydrogen ions than hydroxide ions is acidic*; etc.)
- solution B
- explanation of what makes a solution basic (e.g., *A solution with a higher concentration of hydroxide ions than hydrogen ions is basic*; etc.)

Unit Test B
Key Concepts
1. D   5. A   9. C
2. D   6. A   10. B
3. C   7. D   11. C
4. B   8. A   12. C

1. D

A is incorrect because bleach is a stronger base than sea water and therefore has a higher hydroxide ion concentration.

B is incorrect because orange juice is a stronger acid than pure rain and therefore has a higher hydrogen ion concentration.

C is incorrect because egg whites are a weaker base than household lye and therefore have a lower hydroxide ion concentration.

D is correct because the more acidic a substance is, the greater its hydrogen ion concentration is.

2. D

A is incorrect because a salt will not reduce the pH of ocean water.

B is incorrect because a base will raise the pH of ocean water.

C is incorrect because dirt particles will not lower the pH of ocean water.

D is correct because an acid will lower the pH of ocean water.

3. C

A is incorrect because the solute in a solution will not be separated by filtering.

B is incorrect because water can pass through a filter.

C is correct because filtering will not separate the particles in a solution.

D is incorrect because a filter will not cause the parts of a solution to change chemically.

4. B

A is incorrect because the change is greater than this, as

can be seen by comparing the solubilities at both temperatures.

B is correct because the solubility rises by about 20 grams per 100 g H₂O as the temperature is increased.

C is incorrect because the change is less than this, as can be seen by comparing the solubilities at both temperatures.

D is incorrect because the change is less than this, as can be seen by comparing the solubilities at both temperatures.

5. A

A is correct because the majority of acid rain forms when emissions from burning fossil fuel, such as gasoline, produce sulfuric and nitric acids that combine with water in the atmosphere.

B is incorrect because spraying chemical pesticides does not typically release sulfuric and nitric acids, the primary components of acid rain, into the atmosphere.

C is incorrect because smokestack filters would prevent carbon emissions, a primary cause of acid rain, from entering the atmosphere.

D is incorrect because acid rain is unrelated to the levels of ozone in the atmosphere.

6. A

A is correct because lemons contain citric acid. The acid

and base took part in a neutralization reaction.

B is incorrect because bases do not change into acids.

C is incorrect because the phenolphthalein would continue to indicate the presence of a base.

D is incorrect because both the base and the phenolphthalein it contained were injected into the lemon.

7. D

A is incorrect because formic acid is used for defense.

B is incorrect because ants do not eat formic acid.

C is incorrect because formic acid is used for defense.

D is correct because ants use formic acid in defense from predators.

8. A

A is correct because the particles in a colloid are spread throughout. They are larger than those in a solution but smaller than those in a suspension.

B is incorrect because the particles in solutions are smaller than the particles in either a solution or a suspension.

C is incorrect because the particles in solutions are smaller than the particles in either a solution or a suspension.

D is incorrect because the particles in suspensions are larger than the particles in either a solution or a colloid.

9. C

A is incorrect because the concentration of hydrogen ions in a solution increases tenfold with each whole number drop in pH.

B is incorrect because the concentration of hydrogen ions in a solution increases tenfold with each whole number drop in pH.

C is correct because every unit change on the pH scale is equal to a tenfold change in the hydrogen ion concentration.

D is incorrect because the difference in hydrogen ion concentration between each unit on the pH scale is tenfold.

10. B

A is incorrect because an acid produces hydrogen ions.

B is correct because bases produce hydroxide ions. Sodium hydroxide breaks up into sodium ions and hydroxide ions.

C is incorrect because a salt is formed when an acid reacts with a base, not with water.

D is incorrect because the base dissolves in water. It does not produce more water.

11. C

A is incorrect because the concentration is found by dividing 25 by 125, and then multiplying the quotient by 100%.

B is incorrect because the concentration relates the amount of solute to the total

Answer Key

amount of solution. Therefore, the mass should be divided by the volume, not the volume by the mass.

C is correct because 25 divided by 125 equals 0.20, and 0.20 multiplied by 100% equals 20%.

D is incorrect because the concentration relates the amount of solute to the total amount of solution. Therefore, the mass should be divided by the volume, not added to the volume.

12. **C**

A is incorrect because water is not formed in this process. It is formed when an acid reacts with a base.

B is incorrect because oxygen gas is not produced during this reaction.

C is correct because a reaction between an acid and a metal produces an ionic compound and hydrogen gas.

D is incorrect because hydroxide ions are donated to solutions by bases not acids.

Critical Thinking
13.

- definition of solubility based on the example (e.g., *In this solution, 357 grams of sodium chloride will dissolve in water at this temperature and pressure*; etc.)

- explanation relating solubility to saturation (e.g., *The solution is saturated when 357 g of sodium chloride have been added, so any additional solute will fall to the bottom of the container*; etc.)

Extended Response
14.

- any acid (e.g., *hydrochloric acid*; etc.)

- any base (e.g., *sodium hydroxide*; etc.)

- explanation of what makes a solution acidic or basic (e.g., *Solution A is acidic and was formed when an acid dissolved, whereas Solution B is basic and formed when a base dissolved*; etc.)

- explanation of a neutralization reaction (e.g., *An acid and a base mix in a neutralization reaction to form a salt and water*; etc.)

End-of-Module Test

1. A   11. C   21. A
2. B   12. A   22. A
3. A   13. D   23. B
4. C   14. D   24. A
5. C   15. D   25. B
6. B   16. B   26. D
7. A   17. B   27. A
8. C   18. C   28. A
9. C   19. C   29. D
10. B   20. C   30. B

1. **A See Unit 1, Lesson 3**

A is correct because decaying organic matter involves a change in chemical composition and gives off an odor. This process is a chemical change.

B is incorrect because melting is a state change, which is a physical change rather than a chemical change.

C is incorrect because the wind is making a physical change, not a chemical change, in the cloud's shape.

D is incorrect because breaking into pieces is a physical change in the shape of the statue, not a chemical change.

2. **B See Unit 3, Lesson 1**

A is incorrect because the Bohr model built on the Thomson model and included protons and electrons.

B is correct because the Bohr model showed a nucleus of protons and neutrons surrounded by electrons traveling in specific energy levels.

C is incorrect because the Dalton model, which preceded Thomson's model, showed atoms as solid spheres.

D is incorrect because neutrons are in the nucleus rather than in the region surrounding the nucleus.

3. **A See Unit 3, Lesson 1**

A is correct because atoms combine in many different ways to make up the substances people encounter every day.

B is incorrect because molecules are made up of atoms.

C is incorrect because atoms are not destroyed upon normal heating and cooling.

D is incorrect because atoms are much too small to see using only a hand lens.

4. **C See Unit 2, Lesson 3**

A is incorrect because energy is transferred when matter changes state.

Answer Key

B is incorrect because when energy is transferred into water, the water gets warmer and does not freeze.

C is correct because energy is transferred from the water to the air, which causes water molecules to move less rapidly and freeze.

D is incorrect because energy is not transferred into the ground.

5. C See Unit 1, Lesson 1

A is incorrect because this reading includes the value on only one of the three beams.

B is incorrect because all three beams must be read to find the correct mass.

C is correct because the arrows point to 200, 50, and 4.

D is incorrect because the top arrow points to 4, and not to 4.5.

6. B See Unit 3, Lesson 2

A is incorrect because elements are arranged on the periodic table by increasing atomic number. If zinc is directly to the left of gallium, its atomic number must be 30, or 1 less than gallium's.

B is correct because elements are arranged on the periodic table by increasing atomic number. If zinc is directly to the left of gallium, its atomic number must be 30, or 1 less than gallium's.

C is incorrect because elements are arranged on the periodic table by increasing atomic number. If zinc is directly to the left of gallium, its atomic number must be 30, or 1 less than gallium's.

D is incorrect because elements are arranged on the periodic table by increasing atomic number. If zinc is directly to the left of gallium, its atomic number must be 30, or 1 less than gallium's.

7. A See Unit 3, Lesson 3

A is correct because a full outermost energy level for fluorine will consist of 8 electrons.

B is incorrect because fluorine needs 1 more electron to fill its outermost energy level.

C is incorrect because fluorine needs 1 more electron to fill its outermost energy level.

D is incorrect because fluorine needs 1 more electron to fill its outermost energy level.

8. C See Unit 1, Lesson 6

A is incorrect because matter does not lose particles during a change of state.

B is incorrect because matter does not gain particles during a change of state.

C is correct because mass is conserved for all changes of state.

D is incorrect because matter does not lose particles during a change of state.

9. C See Unit 5, Lesson 2

A is incorrect because acids turn blue litmus paper red.

B is incorrect because bases feel slippery and taste bitter.

C is correct because acids donate hydrogen ions, which bond to water molecules to form hydrogen ions when dissolved in water.

D is incorrect because salts are ionic compounds formed when a metal atom replaces the hydrogen in an acid.

10. B See Unit 4, Lesson 2

A is incorrect because water does not contain carbon and is not an organic compound.

B is correct because the scientist is studying polymers, and starch is an example of a polymer.

C is incorrect because the acids that form acid rain are not long-chain carbon molecules.

D is incorrect because radioactive tracers are unstable isotopes of some elements and not long-chain carbon molecules.

11. C See Unit 2, Lesson 1

A is incorrect because chemical energy is stored in the bonds that hold matter together, and this form of energy will not change during compression.

B is incorrect because electrical charges are not transferred to the object during compression.

C is correct because the spring gains potential energy during compression, which is part of its total mechanical energy.

D is incorrect because electromagnetic energy is transferred as waves, such as light.

12. A See Unit 1, Lesson 5

A is correct because the particles in solids are

constantly vibrating but cannot move past one another.

B is incorrect because the particles in liquids can move past one another, but the particles in solids cannot.

C is incorrect because the particles in solids are constantly vibrating.

D is incorrect because the particles in solids are constantly vibrating, even though they are locked in place.

13. D See Unit 2, Lesson 2

A is incorrect because the kinetic theory of matter states that all of the particles that make up matter are constantly in motion.

B is incorrect because the kinetic theory of matter states that all of the particles that make up matter are constantly in motion.

C is incorrect because the kinetic theory of matter states that all of the particles that make up matter are constantly in motion.

D is correct because the kinetic theory of matter states that all of the particles that make up matter are constantly in motion.

14. D See Unit 2, Lesson 1

A is incorrect because the battery converts chemical energy to electrical energy.

B is incorrect because the battery converts chemical energy to electrical energy.

C is incorrect because the spinning blades have kinetic energy.

D is correct because the battery converts chemical energy to electrical energy, and the electrical energy converts to kinetic energy, which makes the blades spin.

15. D See Unit 1, Lesson 5

A is incorrect because the gas state is not the only state to contain moving particles.

B is incorrect because the particles in gases are also constantly in motion.

C is incorrect because the particles in solids are also constantly in motion.

D is correct because the particles in all states of matter are in constant motion, although the range of motion may vary.

16. B See Unit 3, Lesson 2

A is incorrect because 20 is the atomic number, which is the number of protons in one atom of the element.

B is correct because a chemical symbol is a letter or letter combination that stands for the element.

C is incorrect because 40.078 is the average atomic mass of the element.

D is incorrect because it is the name of the element.

17. B See Unit 5, Lesson 1

A is incorrect because both compounds dissolve in water, not in each other.

B is correct because solubility is the ability of one substance to dissolve in another at a given temperature and pressure.

C is incorrect because solubility does not measure the pH (acidity or basicity) of a compound.

D is incorrect because solubility does not compare the densities of the compounds.

18. C See Unit 2, Lesson 4

A is incorrect because solar energy comes from the sun.

B is incorrect because nuclear energy is released when the centers of atoms are split apart.

C is correct because heat is produced below Earth's surface, and this heat can be used to produce electricity in some places.

D is incorrect because hydroelectric energy involves using falling water to generate electricity.

19. C See Unit 2, Lesson 2

A is incorrect because the temperature of the water near the door is 42°C, which is higher than the temperature of 38 °C in the back corner. The particles move faster, not slower, here.

B is incorrect because the temperature of the water by the window is 45°C, which is higher than the temperature of 38 °C in the back corner. The particles move faster, not slower, here.

C is correct because the temperature of the water in the back corner is 38°C, and this is the lowest of the four measurements. A lower temperature means the particles of water have less energy and move slowest.

D is incorrect because the temperature of the water on the teacher's desk is 43°C, which is higher than the temperature of 38 °C in the back corner. The particles move faster, not slower, here.

20. **C See Unit 1, Lesson 6**

A is incorrect because matter loses energy and particles slow down when matter freezes.

B is incorrect because deposition occurs when a sample of matter loses kinetic energy.

C is correct because matter must absorb energy in order for evaporation to occur.

D is incorrect because condensation occurs when matter cools and loses energy.

21. **A See Unit 1, Lesson 1**

A is correct because the mass (8.1 g) divided by the density (2.7 g/cm^3) equals the volume.

B is incorrect because 2.7 g is the mass of 1 cm^3 of aluminum.

C is incorrect because 0.33 cm^3 the value of the density divided by the mass; volume is calculated by dividing the mass by the density.

D is incorrect because 21.9 is the numerical value obtained by multiplying the mass and density; volume is calculated by dividing the mass by the density.

22. **A See Unit 1, Lesson 3**

A is correct because separation by magnetism, boiling point, and filtration based on particle size are all examples of physical changes.

B is incorrect because separation by magnetism, boiling point, and filtration based on particle size are all examples of physical changes, not chemical changes.

C is incorrect because Austin does not use both chemical and physical methods to separate the mixture.

D is incorrect because Austin does use either chemical or physical methods to separate the mixture.

23. **B See Unit 3, Lesson 4**

A is incorrect because no atoms are destroyed when a water molecule forms.

B is correct because the atoms in a water molecule form covalent bonds.

C is incorrect because hydrogen atoms do not give up electrons in the process shown.

D is incorrect because the oxygen does not give up electrons to become stable.

24. **A See Unit 5, Lesson 3**

A is correct because a higher pH indicates that the blood is more basic.

B is incorrect because water does not raise the pH of the blood.

C is incorrect because salts are neutral and do not raise the pH of blood.

D is incorrect because an increased level of acids would lower pH.

25. **B See Unit 1, Lesson 2**

A is incorrect because physical properties do not behave the same for all matter under the same conditions. Instead, physical properties vary for different kinds of matter.

B is correct because physical properties can be observed without changing the identity of a substance.

C is incorrect because physical properties cannot be observed by seeing how a substance reacts with other substances. Instead, physical properties can be observed without changing the composition of a substance.

D is incorrect because physical properties do not cause atoms or molecules to change. Instead, physical properties can be observed without changing the composition of a substance.

26. **D See Unit 1, Lesson 4**

A is incorrect because carbon is a black solid, and sucrose is a white, crystalline solid.

B is incorrect because oxygen is a colorless gas, and sucrose is a white, crystalline solid.

Answer Key

C is incorrect because hydrogen is a colorless gas, and sucrose is a white, crystalline solid.

D is correct because the properties of a compound differ from the properties of the elements that make it up.

27. A See Unit 1, Lesson 2

A is correct because tarnishing is a chemical property.

B is incorrect because color is a physical property, not a chemical property.

C is incorrect because luster is a physical property, not a chemical property.

D is incorrect because melting point is a physical property, not a chemical property.

28. A See Unit 4, Lesson 1

A is correct because atoms are rearranged during chemical reactions.

B is incorrect because atoms are not created during a chemical reaction.

C is incorrect because atoms are not destroyed during a chemical reaction.

D is incorrect because atoms cannot stay in the same arrangement if new substances are formed.

29. D See Unit 4, Lesson 3

A is incorrect because the atom emitted an alpha particle, which consists of 2 protons and 2 neutrons.

B is incorrect because the atom emitted an alpha particle, which consists of 2 protons and 2 neutrons.

C is incorrect because the atom emitted an alpha particle, which has a mass number of 4.

D is correct because the atom emitted an alpha particle, which has a mass number of 4. The mass number of the atom decreases by 4.

30. B See Unit 4, Lesson 1

A is incorrect because if energy were absorbed, the amount of energy of the products would be higher than the energy of the reactants, not lower.

B is correct because the amount of energy of the products is lower than the amount of energy of the reactants, so this is an exothermic reaction.

C is incorrect because energy is not destroyed during a chemical reaction.

D is incorrect because energy is required for the reaction to proceed.